Aquaculture Technology in Developing Countries

Aquaculture technology has been evolving rapidly over the last two decades, led by an increasingly skilled cadre of researchers in developing countries. Rather than copying, or adapting work done in industrialized countries to their situations, these scientists are moving aquaculture research out of the box to explore species and production systems relevant to their natural resources, economies and social institutions.

Studies from India, Latin America, the Middle East and Africa are highlighted in this collection of papers, covering the entire gamut of aquaculture science from comparison of tilapia breeds, novel feed ingredients for indigenous species, improving disease resistance, water-use efficiency, traditional farming systems, spatial planning and economics. More than a how-to book, this volume introduces the researchers and institutions leading the development of aquaculture as it expands into new frontiers.

This book was based on a special issue of the *Journal of Applied Aquaculture*.

Randall E. Brummett is the Senior Aquaculture Specialist at the World Bank and the Academic Editor of the *Journal of Applied Aquaculture*. He spent 30 years doing research for aquaculture development in Africa and the Middle East.

Aquaculture Technology in Developing Countries

Edited by
Randall E. Brummett

LONDON AND NEW YORK

First published in paperback 2024

First published 2014
by Routledge
4 Park Square, Milton Park, Abingdon, Oxon OX14 4RN

and by Routledge
605 Third Avenue, New York, NY 10158

Routledge is an imprint of the Taylor & Francis Group, an informa business

© 2014, 2024 Taylor & Francis

All rights reserved. No part of this book may be reprinted or reproduced or utilised in any form or by any electronic, mechanical, or other means, now known or hereafter invented, including photocopying and recording, or in any information storage or retrieval system, without permission in writing from the publishers.

Trademark notice: Product or corporate names may be trademarks or registered trademarks, and are used only for identification and explanation without intent to infringe.

Publisher's Note
The publisher accepts responsibility for any inconsistencies that may have arisen during the conversion of this book from journal articles to book chapters, namely the possible inclusion of journal terminology.

Disclaimer
Every effort has been made to contact copyright holders for their permission to reprint material in this book. The publishers would be grateful to hear from any copyright holder who is not here acknowledged and will undertake to rectify any errors or omissions in future editions of this book.

British Library Cataloguing-in-Publication Data
A catalogue record for this book is available from the British Library

ISBN: 978-0-415-82778-2 (hbk)
ISBN: 978-1-03-292583-7 (pbk)
ISBN: 978-1-315-54029-0 (ebk)

DOI: 10.4324/9781315540290

Typeset in Garamond
by Taylor & Francis Books

Contents

Citation Information ix

1. Overview
   R. E. Brummett 1

Africa (policy and production systems)

2. Constraints in Accessing Credit Facilities for Rural Areas: The Case of Fish Farmers in Rural Morogoro, Tanzania
   Kitojo Wetengere and Viscal Kihongo 5

3. The Traditional Whedo Aquaculture System in Northern Benin
   Melanie E. Hauber, David Bierbach, and Karl Eduard Linsenmair 16

4. Production Parameters and Economics of Small-Scale Tilapia Cage Aquaculture in the Volta Lake, Ghana
   J. K. Ofori, E. K. Abban, A. Y. Karikari, and R. E. Brummett 34

5. Impacts of Aquaculture Development Projects in Western Cameroon
   Victor Pouomogne, Randall E. Brummett, and M. Gatchouko 49

6. Rearing Rabbits Over Earthen Fish Ponds in Rwanda: Effects on Water and Sediment Quality, Growth, and Production of Nile Tilapia *Oreochromis niloticus*
   Simon Rukera Tabaro, Onisimo Mutanga, Denis Rugege, and Jean-Claude Micha 65

Middle East (water efficiency)

7. Improving Water Use Efficiency in Semi-Arid Regions through Integrated Aquaculture/Agriculture
   Sami Abdul-Rahman, I. Patrick Saoud, Mohammed K. Owaied, Hanafy Holail, Nadim Farajalla, Mustafa Haidar, and Joly Ghanawi 77

8. Growth Performance of Improved (EXCEL) and a Non-Improved Strains of *Oreochromis niloticus* Fry in a Recirculating Tank System in the UAE
   Nowshad M. Rasheed and Ibrahim E. H. Belal 96

9. Use of Underground Brackish Water for Reproduction and Larviculture of Rainbow Trout, *Oncorhynchus mykiss*
   M. Mohammadi, H. Sarsangi, M. Askari, A. Bitaraf, N. Mashaii, F. Rajabipour, and M. Alizadeh ... 103

10. The Use of American Ginseng (*Panax quinquefolium*) in Practical Diets for Nile Tilapia (*Oreochromis niloticus*): Growth Performance and Challenge with *Aeromonas hydrophila*
    Mohsen Abdel-Tawwab ... 112

Latin America (indigenous species)

11. Use of Spray-Dried Blood Meal as an Alternative Protein Source in Pirarucu (*Arapaima gigas*) Diets
    Ricardo Amaral Ribeiro, Rodrigo Otávio de Almeida Ozório, Sónia Maria Gomes Batista, Manoel Pereira-Filho, Eduardo Akifumi Ono, and Rodrigo Roubach ... 123

12. Effect of Density on Growth and Feeding of the Crayfish *Cambarellus montezumae* (Saussure, 1857)
    José Luis Arredondo-Figueroa, Angélica Vásquez-González, Irene de Los A. Barriga-Sosa, Claudia Carmona-Osalde, and Miguel Rodríguez-Serna ... 135

13. Growth of Juvenile Crayfish *Procambarus llamasi* (Villalobos 1955) Fed Different Farm and Aquaculture Commercial Foods
    M. Rodriguez-Serna, C. Carmona-Osalde, and J. L. Arrendondo-Figueroa ... 143

14. Survival of Diploid and Triploid *Rhamdia quelen* Juveniles Under Different Oxygen Concentrations
    Luciano Augusto Weiss and Evoy Zaniboni-Filho ... 152

15. Effect of Stock Density and Ploidy in Jundia, *Rhamdia quelen*, Larvae Performance
    Hirla Fukushima, Jhon Edison Jimenez, Marcos Weingartner, and Evoy Zaniboni-Filho ... 161

Asia (environment)

16. Integrated Fish Farming for Nutritional Security in Eastern Himalayas, India
    B. P. Bhatt, K. M. Bujarbaruah, K. Vinod, and M. Karunakaran ... 171

17. Wastewater Aquaculture by the Mudialy Fisherman's Cooperative Society in Kolkata, West Bengal: An Example of Sustainable Development
    Archana Sengupta, Tapasi Rana, Biswajit Das, and Shamee Bhattacharjee ... 180

18. The Effect of Partial Replacement of Dietary Fishmeal with Fermented Prawn Waste Liquor on Juvenile Sea Bass Growth
    N. M. Nor, Z. Zakaria, M. S. A. Manaf, and M. M. Salleh ... 190

19. Growth Performance and Resistance to *Streptococcus iniae* of Juvenile Nile Tilapia (*Oreochromis niloticus*) Fed Diets Supplemented with GroBiotic-A and Brewtech Dried Brewers Yeast
   Kunthika Vechklang, Chhorn Lim, Surintorn Boonanuntanasarn, Thomas Welker, Samorn Ponchunchuwong, Phillip H. Klesius, and Chokchai Wanapu — 197

20. Accumulation and Clearance of Orally Administered Erythromycin in Adult Nile Tilapia (*Oreochromis niloticus*) and Giant Freshwater Prawn (*Macrobrachium rosenbergii*)
   N. P. Minh, T. B. Lam, N. T. Giao, and N. C. Quan — 213

Index — 221

Citation Information

The following chapters were originally published in various issues of the *Journal of Applied Aquaculture*. When citing this material, please use the original page numbering for each article, as follows:

Chapter 2
Constraints in Accessing Credit Facilities for Rural Areas: The Case of Fish Farmers in Rural Morogoro, Tanzania
Kitojo Wetengere and Viscal Kihongo
Journal of Applied Aquaculture, volume 24, issue 2 (2012) pp. 107-117

Chapter 3
The Traditional Whedo Aquaculture System in Northern Benin
Melanie E. Hauber, David Bierbach, and Karl Eduard Linsenmair
Journal of Applied Aquaculture, volume 23, issue 1 (2012) pp. 67-84

Chapter 4
Production Parameters and Economics of Small-Scale Tilapia Cage Aquaculture in the Volta Lake, Ghana
J. K. Ofori, E. K. Abban, A. Y. Karikari, and R. E. Brummett
Journal of Applied Aquaculture, volume 22, issue 4 (2012) pp. 337-351

Chapter 5
Impacts of Aquaculture Development Projects in Western Cameroon
Victor Pouomogne, Randall E. Brummett, and M. Gatchouko
Journal of Applied Aquaculture, volume 22, issue 2 (2012) pp. 93-108

Chapter 6
Rearing Rabbits Over Earthen Fish Ponds in Rwanda: Effects on Water and Sediment Quality, Growth, and Production of Nile Tilapia Oreochromis niloticus
Simon Rukera Tabaro, Onisimo Mutanga, Denis Rugege, and Jean-Claude Micha
Journal of Applied Aquaculture, volume 24, issue 2 (2012) pp. 170-181

Chapter 7
Improving Water Use Efficiency in Semi-Arid Regions through Integrated Aquaculture/Agriculture
Sami Abdul-Rahman, I. Patrick Saoud, Mohammed K. Owaied, Hanafy Holail, Nadim Farajalla, Mustafa Haidar, and Joly Ghanawi
Journal of Applied Aquaculture, volume 23, issue 3 (2012) pp. 212-230

Chapter 8
Growth Performance of Improved (EXCEL) and a Non-Improved Strains of Oreochromis niloticus *Fry in a Recirculating Tank System in the UAE*
Nowshad M Rasheed and Ibrahim E. H. Belal
Journal of Applied Aquaculture, volume 22, issue 4 (2012) pp. 352-358

Chapter 9
Use of Underground Brackish Water for Reproduction and Larviculture of Rainbow Trout, Oncorhynchus mykiss
M. Mohammadi, H. Sarsangi, M. Askari, A. Bitaraf, N. Mashaii, F. Rajabipour, and M. Alizadeh
Journal of Applied Aquaculture, volume 24, issue 2 (2012) pp. 137-146

Chapter 10
*The Use of American Ginseng (*Panax quinquefolium*) in Practical Diets for Nile Tilapia (*Oreochromis niloticus*): Growth Performance and Challenge with* Aeromonas hydrophila
Mohsen Abdel-Tawwab
Journal of Applied Aquaculture, volume 24, issue 4 (2012) pp. 366-376

Chapter 11
*Use of Spray-Dried Blood Meal as an Alternative Protein Source in Pirarucu (*Arapaima gigas*) Diets*
Ricardo Amaral Ribeiro, Rodrigo Otávio de Almeida Ozório, Sónia Maria Gomes Batista, Manoel Pereira-Filho, Eduardo Akifumi Ono, and Rodrigo Roubach
Journal of Applied Aquaculture, volume 23, issue 3 (2012) pp. 238-249

Chapter 12
Effect of Density on Growth and Feeding of the Crayfish Cambarellus montezumae *(Saussure, 1857)*
José Luis Arredondo-Figueroa, Angélica Vásquez-González, Irene de Los A. Barriga-Sosa, Claudia Carmona-Osalde, and Miguel Rodríguez-Serna
Journal of Applied Aquaculture, volume 22, issue 1 (2010) pp. 66-73

Chapter 13
Growth of Juvenile Crayfish Procambarus llamasi *(Villalobos 1955) Fed Different Farm and Aquaculture Commercial Foods*
M. Rodriguez-Serna, C. Carmona-Osalde, and J. L. Arrendondo-Figueroa
Journal of Applied Aquaculture, volume 22, issue 2 (2010) pp. 140-148

Chapter 14
Survival of Diploid and Triploid Rhamdia quelen *Juveniles Under Different Oxygen Concentrations*
Luciano Augusto Weiss and Evoy Zaniboni-Filho
Journal of Applied Aquaculture, volume 22, issue 1 (2010) pp. 30-38

Chapter 15
Effect of Stock Density and Ploidy in Jundia, Rhamdia quelen, *Larvae Performance*
Hirla Fukushima, Jhon Edison Jimenez, Marcos Weingartner, and
Evoy Zaniboni-Filho
Journal of Applied Aquaculture, volume 23, issue 2 (2011) pp. 147-156

Chapter 16
Integrated Fish Farming for Nutritional Security in Eastern Himalayas, India
B. P. Bhatt, K. M. Bujarbaruah, K. Vinod, and M. Karunakaran
Journal of Applied Aquaculture, volume 23, issue 2 (2012) pp. 157-165

Chapter 17
Wastewater Aquaculture by the Mudialy Fisherman's Cooperative Society in Kolkata, West Bengal: An Example of Sustainable Development
Archana Sengupta, Tapasi Rana, Biswajit Das, and Shamee Bhattacharjee
Journal of Applied Aquaculture, volume 24, issue 2 (2012) pp. 137-146

Chapter 18
The Effect of Partial Replacement of Dietary Fishmeal with Fermented Prawn Waste Liquor on Juvenile Sea Bass Growth
N. M. Nor, Z. Zakaria, M. S. A. Manaf, and M. M. Salleh
Journal of Applied Aquaculture, volume 23, issue 1 (2011) pp. 51-57

Chapter 19
Growth Performance and Resistance to Streptococcus iniae *of Juvenile Nile Tilapia* (Oreochromis niloticus) *Fed Diets Supplemented with GroBiotic-A and Brewtech Dried Brewers Yeast*
Kunthika Vechklang, Chhorn Lim, Surintorn Boonanuntanasarn, Thomas Welker, Samorn Ponchunchuwong, Phillip H. Klesius, and Chokchai Wanapu
Journal of Applied Aquaculture, volume 24, issue 3 (2011) pp. 183-198

Chapter 20
Accumulation and Clearance of Orally Administered Erythromycin in Adult Nile Tilapia (Oreochromis niloticus) *and Giant Freshwater Prawn* (Macrobrachium rosenbergii)
N. P. Minh, T. B. Lam, N. T. Giao, and N. C. Quan
Journal of Applied Aquaculture, volume 25, issue 1 (2013) pp. 1-8

Overview

R. E. Brummett

Aquaculture as we know it in the early 21st century is a consolidation of more or less independent experiences. Pharonic carvings indicate that the Egyptians were cultivating fish at least 2500 years ago. The Chinese claim to have been growing fish for centuries. The Romans had fishponds (*piscinae*). In the 14th century, the emperor Charles IV ordered all towns to build fish ponds to produce food, enhance the local environment and protect watersheds. Paleolithic Hawaiian Islanders isolated embayments for rearing fish in the sea. Whatever the original objective of these aquaculture initiatives, from each evolved a set of concepts that until quite recently, strongly influenced how aquaculture interacted with local society and the environment.

One could argue that modern, global aquaculture arose from these different local traditions only in the second half of the 20th century. Aquaculture began to gain critical mass and make noticeable contributions to fish supply in the 1950's (Figure 1). Scientific evaluation of the integrated agriculture-aquaculture systems that had evolved in China began in the 1960's. The publication of two books in the 1970's, Traité de Pisciculture, Fourth Edition (Huet 1970) and Aquaculture: The Farming and Husbandry of Freshwater and Marine Organisms (Bardach et al. 1972) for the first time brought together for analysis and comparison the range of global aquaculture experiences. The World Aquaculture Society and the European Aquaculture Society were formed in the 1970's. The journal Aquaculture began in 1972. The Food and Agriculture Organization of the United Nations (FAO) began separate reporting of aquaculture statistics from capture fisheries statistics in the early 1980's.

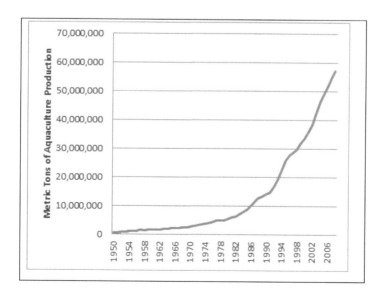

FIGURE 1 The nearly logarithmic expansion of global aquaculture was fueled by a close working arrangement between scientists and investors starting in the 1950's.

These initiatives consolidated, or represented a consolidation of, the wide array of aquaculture systems that had grown in isolation from each other, the international research and development (R&D) community that has worked together with industry to solve problems and produce average annual industrial growth rates of about 10% over the last 15 years.

Aquaculture is predominantly an affair of less-industrialized economies (Figure 2). Despite the dominance of, particularly, Asia in terms of production and consumption, the technology that drives productivity and efficiency is derived from scientific research that is largely conducted and published in Europe and North America. Less than 20% of the world's population lives in the countries of the Organization for Economic Cooperation and Development (OECD) but out of the total of 139 257 engineering and technology papers published in 2007, almost 60% came from OECD countries (UNESCO 2010). Of the 47 574 patents granted by the Triad patent offices (the US Patents and Trademark Office, European Patent Office and Japanese Patent Office) in 2006, 96% were to researchers in the OECD.

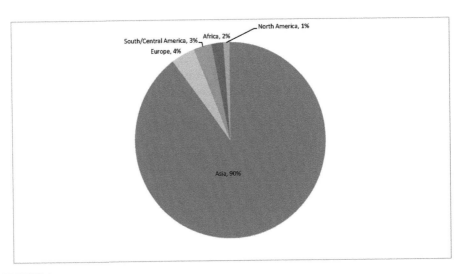

FIGURE 2

In some cases, research and development are regarded as luxuries for industrialized economies and deprioritized in the budgets of less developed countries (LDCs). Without an indigenous and innovative scientific community to guide the development of new technology and the adaptation of old technology, less developed regions will never catch up with already industrialized ones. We can only imagine what could be achieved with the elimination of this waste of 80% of the world's brainpower.

Aquaculture science in developing countries is largely and properly targeted towards the resolution of local problems. Even in industrialized countries, this kind of work is seldom of interest to scientific journals that seek to push theoretical boundaries.

In applied science, however, theory is only useful to the extent that it guides practical problem solving. Because it focuses on a field where LDCs have a clear advantage, aquaculture research can be a leader in over-turning the perception that science is the business of the already wealthy. The Journal of Applied Aquaculture has over recent years focused attention on the need to enhance the profile of practical aquaculture scientists working to solve real problems in the context of less developed economies by getting their research out into the public domain where it can be used to positive by other applied aquaculture scientists.

In this volume, we have brought together some of the outputs of this strategy from over the past three years. Papers from practical aquaculture scientists working in Latin America, Asia, the Middle East and Africa reflect the relative level of development of the sector in these regions and reveal the breadth of analytical, basic and applied research that is generating local solutions for global aquaculture.

As one might expect from the struggling aquaculture sector on the mother continent, papers from Africa focus on development policy and basic production systems, with an emphasis on small-scale aquaculture for rural food security. Herbert Ssegane in Uganda developed a geospatial model to guide local government prioritization of aquaculture in various parts of the country. Kitojo Wetengere and Viscal Kihongo in Tanzania examine the availability of credit that would allow smallholders to ramp up production to meaningful scale and finds that the rural poor cannot get credit because their production systems are too risky. Support to extension services and improved business planning are recommended as means to improve system reliability and profitability, thus making aquaculture farms more bankable. An aquaculture team at Cameroon's Humid Forest Ecoregional Center explores the context of small-scale aquaculture development and finds that, like in Tanzania, quality of technical support to the aquaculture sector is problematic, but credit is less of a constraint than reliable access to feeds and good fingerlings. Overall, these constraints have constrained growth and prevented small-scale aquaculture from having any noticeable impact on rural poverty.

Focusing on African production systems, Melanie Hauber studied the dynamics and potential for improvement of the traditional *whedo* aquaculture system in Benin, concluding that concentrating technological innovation to enhance existing production systems would be more efficient than introducing foreign technology. Simon Tabero in Rwanda explores the productivity of integrated fish-rabbit systems, while Ofori *et al.* describe the economics of growing tilapia in cages in the Volta Lake, Ghana.

Water is a major constraint to aquaculture development in the Middle East, as reflected by contributions from this region. In Lebanon, Imad Saoud's team at the American University of Beirut has been studying mixed fish and row crop production systems that can increase water use efficiency. In the United Arab Emirates, Nowshad Rasheed and Ibrahim Belal show how selective breeding can improve fish performance and water use efficiency in recirculating aquaculture systems. In Iran, Mohammadi *et al.* showed how underground brackishwater of limited use for other food production systems, can successfully be used in trout hatcheries.

The aquaculture sector in Latin America and the Caribbean is growing and the search is on for aquaculture candidates from among the thousands of indigenous fish species in the region. The giant pirarucu from the Amazon performs well in culture, but demands very high protein feeds, so a team in Brazil looked at blood meal as an alternative. Already produced in some quantities, variability in growth performance in jundia catfish lead Luciano Augusto Weiss and Evoy Zaniboni-Filho at the Federal University of Santa Catarina to test the feasibility of using triploidy technology to improve consistency. In Mexico, aquaculture is seen not only as a means of growing aquatic animals for food, but also for restocking and a group led by José Luis Arredondo-Figueroa at the Autonomous University of Metropolitan Iztapalapa is developing culture techniques for the indigenous acocil crayfish.

Coping with over-stress land and water resources is a leading theme in contributions from crowded Asia. Indian scientists in the Eastern Himalayas are working to develop more efficient integrated farming systems that combine crops, land animals and fish. Archana Sengupta from the Vivekananda Institute of Medical Sciences describes how fishers in heavily polluted Kolkata are using aquaculture to make money while improving local water quality. In Malaysia, Zainoha Zakaria leads a team that is looking to replace fishmeal with prawn processing wastes, while in Vietnam a team at Stapimex, one of the country's leading seafood companies, establishes guidelines for the use of antibiotics in prawn culture.

Taken together, this selection of papers from the Journal of Applied Aquaculture exhibits the range and sophistication of the research being carried out to support the continued expansion of the world's fastest growing food production sector. The editorial board and publisher of the journal will continue to work with these scientists as they push the boundaries, and make their good work better available to the global aquaculture community.

References

Bardach, J.E., J.H. Ryther and W.O. McLarney. 1972. *Aquaculture: the farming and husbandry of freshwater and marine organisms*. Wiley-Interscience, New York, USA.

FAO. 2012. *State of world fisheries and aquaculture*. Food and Agriculture Organization of the United Nations, Rome.

Huet, M. 1970. *Traité de Pisciculture* (4th Edition). Editions Ch. De Wyngaert, Brussels, Belgium.

UNESCO. 2010. *UNESCO science report; the current status of science around the world*. United Nations Educational, Scientific and Cultural Organization, Paris.

Constraints in Accessing Credit Facilities for Rural Areas: The Case of Fish Farmers in Rural Morogoro, Tanzania

KITOJO WETENGERE[1] and VISCAL KIHONGO[2]
[1]*Centre for Foreign Relations, Dar es Salaam, Tanzania*
[2]*Institute of Social Work, Dar es Salaam, Tanzania*

This study examined constraints in accessing credit facilities for fish farmers in rural Morogoro, Tanzania. Data was collected using techniques of participatory rural appraisal and secondary information sources. A descriptive statistics method was used to report findings, and data were validated by mean percentages and group consensus. This result revealed that farmers earned an average income per capita of Tshs. 56,666 (US$38) per year. Due to low profitability and high risk, most farmers were reluctant to invest their meager income in fish farming. Only 8% of respondent's accessed credit facilities. Constraints to credit access included lack of information, unfavorable terms, lack of support services, and illiteracy. Research should focus on how to improve the profitability and reduce the risk of fish farming to attract bank lending in the industry.

INTRODUCTION

Poverty in Tanzania is overwhelmingly rural (United Republic of Tanzania [URT] 2005; Tanzania National Bureau of Statistics [TNBS] 2009). Poverty is

---

Mrs. Lilian Wetengere is gratefully acknowledged for her assistance in collecting data for this paper. The authors are sincerely grateful to the anonymous reviewers for their helpful comments regarding the drafts of this study. The final content of this article, however, is the sole responsibility of the authors.

highest among households that depend on agriculture (URT 2005). In rural Morogoro, poverty is manifested by people's inability to produce enough food and income to meet household needs (ALCOM 1994). Farmers need to increase farm productivity in order to improve household food and income security (Wetengere 2010a). To meet rapidly growing food demand and raise rural incomes, farmers must intensify their farm production system (Delgado et al. 2003; Wetengere 2008b, 2010b).

Farmed fish is one of the high-value crops that can be intensified to meet household needs (Wetengere 2008b, 2010b). Despite the high potential for fish farming, a low level of technology has been practiced in Morogoro (Wetengere 2008b, 2010b), characterized by small ponds likened to holes, low stocking density, poor-quality seed, and low inputs in terms of management, labor, feeds, and fertilizers (Wetengere 2008a, 2008b, 2010b). Limited capital and/or lack of access to credit to purchase inputs constrain intensification (Kudi et al. 2009; Mikalitsa 2010; Wetengere 2008b, 2010b).

Although the subject of credit for rural development has been widely examined by, for example, Masawe (1994), Shastri (2009), Kudi et al. (2009), Fansoranti (2010), and Mikalitsa (2010), there are few studies (Johnson 1993; ALCOM 1994; FAO 2000; Wetengere 2008b) that have investigated the availability of credit for development of rural fish farming. The objective of this study, therefore, was to make a thorough investigation of constraints in accessing credit facilities for fish farming in rural Morogoro, Tanzania.

METHODS

The data used in this study are part of a market survey conducted in April–May 2006 in 24 selected villages in the Morogoro region of Tanzania. A field survey design that focuses on individual farmers as the unit of analysis was employed. This method is capable of capturing existing perceptions, attitudes, behaviors, and values of individuals within a household (Mugenda & Mugenda 1999).

The study population consisted of fish farming adopters (women and men). From each village list, a systematic random sampling approach was used to select the respondents in order to avoid conscious or unconscious biases in the selection of sampled households and to ensure that the selected sample was representative of the entire population. In total, 217 farmers were selected for interview from among 580 fish farmers. Given the nature and diversity of the study, a large sample was required to reduce sampling errors (Mugenda & Mugenda 1999). From the village list, additional names were identified for replacement in cases where the respondents selected were not available.

The instruments used for data collection were a structured questionnaire, researcher observation, and participatory rural appraisal (PRA) conducted in each village to solicit information on constraints in assessing credit facilities for rural farmers. Researcher observation was used to corroborate questionnaire responses. In PRA group meetings, questions were discussed and points arrived at after reaching consensus. Three research questions were used to guide this study:

- Do fish farmers reinvest cash income generated from on-farm and off-farm activities in fish farming?
- What are the main sources of credit in the study area?
- What constraints do farmers face in accessing credit for fish farming?

Data analysis was based on descriptive statistics compiled and calculated in the Statistical Package for Social Sciences (SPSS) software.

RESULTS AND DISCUSSIONS

Table 1 presents demographic characteristics of the 217 respondents sampled from the study area. Male respondents were the majority. Most

TABLE 1 Demographic characteristics of the sampled population

Respondents characteristics		Sample size by sex		Total sample n = 217 (%)
		Male n = 168 (%)	Female n = 49 (%)	
Gender/sex		77	23	100
Main occupation	Full time farmer	43	35	42
	Farmer and trader	48	63	51
	Farmer and employee	6	2	5
	Others	3	0	2
Education level	No formal education	9	16	11
	Less than Standard 7	17	16	17
	Standard 7	63	64	64
	Secondary and post secondary	11	4	8
Age	≤30 years	35	39	36
	31–50 years	37	47	39
	≥51 years	28	14	25
Total annual cash income	Income of Tshs. 200,000* and below	44	55	47
	Income between Tshs. 200,001–500,000*	38	31	36
	Income of Tshs. 500,001* and above	18	14	17

*1 US$ = Tshs 1,500.

households do farming as a major livelihood earning activity, but 58% of respondents indicated that farming was not the only household activity. About 51% of respondents derived their livelihood from farming and trading, 42% from full-time farming, 5% by a combination of farming and working on other farms, and 2% from non-farm activities. While more women were doing farming and trading, more men were full-time farmers. The percentage of full-time farmers in Morogoro is lower than the national average of 63% (TNBS 2002). The main type of business/trading in the study area is brewing. Other businesses include small shops selling of timber, charcoal, bricks, and crops.

About 47% of respondents earned a total annual cash income of Tshs. 200,000 (US$134) and below, 36% earned income of Tshs. 200,001–500,000 (US$134–$333), and 17% earned income of Tshs. 500,001 (US$334) and above. On average, women were poorer than men. An average cash income of Tshs. 327,994 (US$219 annually) was earned in the study area in the 2005/06 farming year (Wetengere 2008b).

On-farm activities contributed 44% of the total income and off-farm, 56%. The average income was Tshs. 327,994 (US$219) per year, which amounted to an average cash income of Tshs. 899 (US$0.60) per day. Based on the average household size of 5.9 people (Wetengere 2009) in the study area, this amount gives an average income per capita of Tshs. 56, 666 (US$38) per year or Tshs. 150 (US$0.10) per day, higher than that reported by Maselle, Masanyiwa, and Namwata (2008) in Bukombe district, which was Tshs. 46,844 (US$31) per year. Data from URT (2007) gives an average income per capita from overall rural areas in Tanzania of 28,418 (US$19). These results suggest that most people in the rural areas live far below the poverty line (Maselle, Masanyiwa, & Namwata 2010).

Investment of cash income earned from other sources (other than borrowing from financial institutions) in fish farming was minimal. In addition, only a small amount of cash income earned from fish farming was re-invested into the activity; instead, most of it was diverted into other activities whose benefits were more certain (Nilsson & Wetengere 1994). Wetengere (2010d) observed that only 18% of respondents re-invested some cash obtained from fish farming to further development of the activity. The amount invested was meager, ranging from Tshs. 10,000 to 50,000 (US$7–$33).

Participants in PRA meetings indicated that farmers were reluctant to invest in fish farming for two basic reasons:

1. Fish farming is considered a low income-generating activity, with unknown and unreliable profitability; fish are mostly grown for home consumption (Wetengere 2010a, 2010c).

2. Fish farming is perceived as a high-risk activity because of: a) high animal predation and human theft; b) high probability of pond flooding; c) poor growth of fish due to overcrowding, poor management, and poor-quality fingerlings; (d) death of fish or fingerlings due to overfertilization, fingerling transport, influx of dirty water during rain, and poor harvesting methods; (e) rotting of fish due to poor preparation and preservation methods; (f) low market and poor marketing channels; and (g) and poisoning with crop sprays.

During this survey (April–May, peak farming season) the researchers observed that most ponds were deserted (overgrown by grass, had transparent water, and low water levels), as most farmers were busy attending their crop farms.

Increasing farm productivity to meet household food and income requirements necessitates intensification of farm activities. Due to low income, however, households find it necessary to borrow from various sources to finance their activities (Kudi et al. 2009). Credit is needed to finance activities such as pond construction and repair, purchase of essential equipment (e.g., nets), and procurement of inputs (seed, feed, fertilizer). Almost all respondents reported needing extra cash income to increase farm production. These results concur with findings by World Bank (1989), which revealed that, due to increased costs of inputs, the demand for rural credit has expanded.

Table 2 shows that, in the past 12 months, only 8% of the sampled population obtained credit from various sources (neighbors/relatives/friends, 7%; government institutions, 1%). None of the respondents obtained credit facilities from formal financial institutions, and none of the credit was for aquaculture. In rural areas like Morogoro, farmers go to their well-off neighbors (e.g., teachers, health officers, politicians, and business persons) for borrowing. Farmers mentioned that the main reasons for borrowing from these groups were because they knew and trusted each other; they were more accessible as they all lived in the same locality and had personal relations. Consequently, it was easy to follow-up repayment of loans. Some

TABLE 2 Sources of credit facilities in the study area

Sources of credit	Male (%) n = 168	Female (%) n = 49	Total (%) n = 217
Nowhere to borrow	90	98	92
Neighbors/relatives/friends	9	2	7
Government institutions	1	0	1
Formal financial institutions	0	0	0

Source: Author's survey results, 2007.

farmers mentioned that the common practice was to borrow cash from these groups and promise to pay back in kind (cash crops) when they harvest their crops. Participants in PRA meetings mentioned that the amount borrowed at any one time was small, ranging between Tshs. 10,000 (US$7) and 100,000 (US$67). This was also an observation by Kudi et al. (2009), who found that most farmers considered the farm credit disbursed to them as too small.

Although the role of credit for rural development has been widely documented (Masawe 1994), it has not reached those who actually need it—the rural poor farmers (Lusindilo et al. 2010).

More female respondents than male respondents had no place to obtain credit. There are a number of socio-economic and cultural barriers such as immobility, illiteracy, low formal education, lack of profitable activities and immovable assets, and high workload that prevent women from accessing formal or informal sources of credit (Spliethoff 1994).

Participants in PRA meetings indicated that the low and unreliable incomes generated by fish farming account for the unwillingness of financial institutions to provide credit facilities to fish farmers (Wetengere 2010e). These results are supported by Quagrainie et al. (2010), who argued that fish farmers in Kenya who sold more tilapia were more likely to utilize credit facilities for their fish operations, implying that farmers who sold more fish for cash were seen by lending institutions as having a better chance of paying back their loans.

Kudi et al. (2009) observed that Nigerian banks were not favorably disposed to lending to agriculture and, when they did, it was usually short-term financing (Aghato 2000). Quagrainie et al. (2010) reported contrasting findings in Western Kenya, where commercial banks showed interest in financing fish farming because a high concentration of farmers in one area reduced administrative costs of managing loans.

From past experience, participants in PRA meetings felt that even if credit facilities were made available, they would benefit only the well-to-do farmers, similar to the observation of Fansoranti (2010), who noted that few poor farmers benefited from credit facilities. Some farmers were of the view that, even if credit were available for smaller-scale fish farming, funds would likely be used for non-aquaculture purposes. Evidence exists in other fish farming projects that tools like wheelbarrows, polythene pipes, spades, etc., provided for fish farming were diverted to other (more) profitable and low-risk activities like vegetable gardening, animal husbandry, and irrigated crops. Similar results were observed by Masawe (1994), who found that, in a tractor credit program in Morogoro, tractors were used for transporting people rather than for farming of crops simply because transportation was more profitable than crop farming. Participants in PRA meeting stressed that if fish farming is to attract lending, its profitability has to be established and risk properly managed (Ridler & Hishamunda 2001).

CONSTRAINTS TO ACCESS TO FORMAL CREDIT INSTITUTIONS

Lack of Information

Consistent with Shastri (2009), who found that rural people are generally unaware of banking policies and credit systems, most of the respondents in Morogoro (66%) said that they did not know if there were organizations within or outside the village that provided credit facilities for rural people. Some participants in PRA group meetings were of the opinion that some of the information on availability of credit was not made public or posted in arcane locations intentionally to avoid and/or reduce competition. It was mentioned for instance that some announcements regarding availability of credit were placed at ward offices that were only visited by the politically connected. Women in particular lacked information on the availability of credit facilities, due mostly to their lack of transportation, high level of illiteracy, low formal education, and heavy workload, i.e., lack of spare time (Mikalista 2010). Losindilo et al. (2010) observed similar results and noted that women with pre-school education knew nothing about loan programs, while those with the post-primary education actually obtained loans. During one of the PRA group discussions on availability of information on credit facilities one woman remarked: "With all the household chores and farm work, how do you expect me to know organizations that provide credit facilities?" And she answered the question herself: "Only if I was contacted directly, otherwise I would be bypassed." Since women were fully occupied with household chores and farm work, they hardly had time to attend to their own health and nutrition needs, let alone time for social activities (Mikalista 2010). Unlike women, men spend less time on household chores and only about six hours on the farm per day, leaving more time to socialize.

Difficult to Access

Thirty-four percent of respondents had obtained information on the availability of credit from various sources (radio, newspapers, friends/relatives, etc.). Participants in group PRA meetings indicated, however, that most farmers did not have access to bank credit for a number of reasons: 1) most banks handling credit funds were located in urban centers 20–100 km away from the study area (Lubwama's [1999] study in Uganda also identified distance of financial institutions from the farmers as a hindrance to credit access.); 2) interest rates are too high for poor farmers, ranging from 19% to 22% in Tanzania; and 3) formal banking institutions demanded collateral in the form of land, house, or title of some immovable assets. Most smallholder farmers in Africa cannot fulfill any of these requirements (Lubwama 1999), and the situation is more difficult for women, who have no rights to ownership of immovable assets, are often illiterate, and lack mobility (Spliethoff 1994).

Makilitsa (2010), working in Kenya, established that men had access to credit from formal banks, moneylenders, and cooperatives, while women's sources of credit were kin/friends and local revolving credit schemes.

Most financial institutions prefer lending to a smaller number of big borrowers to minimize loan administrative costs. In Morogoro, there are many small-scale farmers scattered over a wide area. Such a geographical setting makes loan provision cost ineffective. In addition, bankers tend to consider low-income households a bad risk and so incur exceedingly high monitoring and operation costs to protect their loans.

Lack of Non-Credit Supportive Services

Clargo (2009) found that many people do not have the technical or managerial knowledge needed to run a business. According to Otieno, Nyikal, and Mugivane (2009), non-credit supportive services considered important include training in business plan development and implementation; phased enterprise development and near-term marketing options; sub-sector analysis and appropriate buyer sector selection; quality assurance and standards; use of innovative technologies as mobile phones for timely market information and enhancement of negotiation skills; and product grading, packing, and branding for various buyer sectors (Ledgerwood 1999). Some farmers in Morogoro mentioned that they did not borrow for fear of failing to operate the business profitably due to lack of knowledge and business skills. Farmers mentioned that they hear over the radio and read from newspapers that there is training on business management, but these courses are only conducted in urban centers. Even when done locally, Masawe (1994) found that most of these services were inadequate to facilitate smooth operation of a credit program. Provision of credit particularly to small farmers without support services was a waste of time. Other studies have also shown that the availability of quality support services contribute positively to the success of many credit projects (Masawe 1994; Kudi et al. 2009). Similar results by Thomas and Dennis (2009) revealed that the provision of these services reduces the risk associated with running a project and increases the chance of success.

Illiteracy

Managing credits from formal financial institutions requires at least some knowledge in reading, writing, and calculation. Such knowledge is important in completing credit forms, understanding training courses, operating cash flow analysis, and bookkeeping and saving. This study found that 11% of all respondents were illiterate and therefore would have difficulties in accessing credit from formal financial institutions.

RECOMMENDATIONS

Some policy ramifications emerge from this study. First, since the level of credit use is very low in the study area, there is a need to provide information to rural farmers on the availability and management of credit. Second, the government, lending institutions, and other stakeholders should deal with constraints hindering rural farmers from accessing credit facilities. This would mean finding ways of lending to poor farmers without requiring collateral assets, as has been the case with the Grameen Bank in Bangladesh or Village Community Banks (VICOBA). Third, zones where most fish farms concentrate and where fish farming has a high potential of success must be identified; this would reduce the risk and administrative costs of managing loans to attract bank lending. Finally, more research should be conducted on how to reduce risk and improve the profitability of fish farming to attract lending in the industry.

REFERENCES

Aghato, O. A. 2000. Refinancing scheme for medium and long gestation agricultural projects in Nigeria. Workshop given by the Central Bank of Nigeria.

Aquaculture for Local Community Development Programme (ALCOM). 1994. *Background report on fish farming in Morogoro, Tanzania*. Morogoro, Tanzania: ALCOM.

Clargo, M. 2009. The entrepreneurship cycle. In *Footsteps, vol. 80*. http://tilz.tearfund.org/Publications/Footsteps+71-80/Footsteps+80/The+entrepreneurial+cycle.htm

Delgado, C., N. Wada, M. Rosegrant, S. Meijer, and M. Ahmed. 2003. *Fish to 2020: Supply and demand in changing global markets. WorldFish Center technical report 62*. Washington, DC: International Food Policy Research Institute (IFPRI) and WorldFish Center.

Food and Agriculture Organization of the United Nations (FAO). 2000. *Small ponds make a big difference. Integrating fish and livestock farming*. Rome: FAO.

Johnson, C. 1993. People, fish, and ponds: A study of aquaculture in Malawi. Central and Northern fish farming project, socio-economic survey. Central and Northern Region Fish Farming Project, Malawi, pp. 67.

Kudi T. M., S. B. Odugbo, A. L. Banta, and M. B. Hassan. 2009. Impact of UNDP microfinance programme on poverty alleviation among farmers in selected local government areas of Kaduna State, Nigeria. *International Journal of Sociology and Anthropology* 1(6): 99–103.

Ledgerwood, J. 1999. *Microfinance handbook: An institutional and financial perspective. Sustainable banking with the poor*. Washington, DC: The World Bank.

Losindilo, E., A. S. Mussa, and R. R. J. Akarro. 2010. Some factors that hinder women participation in social, political and economic activities in Tanzania. *Arts and Social Sciences Journal* 1:ASSJ–4.

Lumbwana, F. B. 1999. Socio-economic and gender issues affecting the adoption of conservation tillage practices. In *Conservation tillage with animal traction*, edited by P. G. Kaumbutho and T. E. Simalenga,. Harare, Zimbabwe: Animal Traction Network for Eastern and Southern Africa (ATNESA).

Masawe, J. L. 1994. Agricultural credit as an instrument of rural development in Tanzania: A case study on the credit programme for tractorization of small-scale agriculture in Morogoro Region. *African Study Monographs* 15(4): 211–226.

Maselle, A. E., Z. S. Masanyiwa, and B. M. L. Namwata. 2008. Demographic context and its implications for rural household poverty in Bukombe District, Tanzania. *Rural Planning Journal* 10(1): 99–112.

Maselle, A. E, Z. S. Masanyiwa, and B. M. L. Namwata. 2010. Household socio-economic characteristics and poverty in Tanzania. A case of selected villages in Bukombe district. *The Institute of Social Work Journal* 1(1): 59–76.

Mikalista, S. M. 2010. Gender-specific constraints affecting technology use and household food security in Western Province Kenya. *African Journal of Food Agriculture Nutrition and Development* 10(4): 2,324–2,343.

Mugenda, O., and A. Mugenda. 1999. *Research methods: Quantitative and qualitative approaches.*, Nairobi, Kenya: African Center for Technology Studies (ACTS) Press.

Nilsson, H., and K. Wetengere. 1994. *Adoption and viability criteria for semi-intensive fish farming: A socio-economic study in Ruvuma and Mbeya regions in Tanzania. ALCOM/FAO field document no. 28*. Harare, Zimbabwe: ALCOM

Otieno, P. S., R. A. Nyikal, and F. I. Mugivane. 2009. Non-credit services of group-based financial institutions: Implications for smallholder women's honey income in arid and semi arid lands of Kenya. *African Journal of Agricultural Research* 5(5): 344–347.

Quagrainie, K. K., C. C. Ngugi, and S. Amisah. 2010. Analysis of the use of credit facilities by small-scale fish farmers in Kenya. *Aquaculture International* 18:393–402. doi: 10.1007/s10499-009-9252-8.

Ridler, N., and N. Hishamunda. 2001. *Promotion of sustainable commercial aquaculture in sub-Saharan Africa. Volume 1: Policy framework. FAO Fisheries technical paper no. 408/1*. Rome: FAO.

Shastri, R. K. 2009. Micro finance and poverty reduction in India (A comparative study with Asian countries). *African Journal of Business Management* 3(4): 136–140.

Spliethoff, P. C. 1994. *Fisheries and aquaculture: Operational guidelines for the incorporation of gender in project/programme preparation and design.* Brussels: European Commission Publishing.

Tanzania National Bureau of Statistics (TNBS). 2002. *Household budget survey (HBS) 2000/01*. Dar es Salaamm, Tanzania: TNBS.

Tanzania National Bureau of Statistics (TNBS). 2009. *Household budget survey (HBS) 2007*. Dar es Salaamm, Tanzania: TNBS.

Thomas, S., and R. Dennis. 2009. Starting your own business. In *Footsteps*, vol. 80. http://tilz.tearfund.org/Publications/Footsteps+71-80/Footsteps+80/Starting+your+own+business.htm

United Republic of Tanzania (URT). 2005. *National strategy for growth and reduction of poverty (NSGRP)*. Dar es Salaam, Tanzania: URT Vice President's Office.

United Republic of Tanzania (URT). 2007. *Poverty and human development report 2007*. Dar es Salaam, Tanzania: URT.

Wetengere, K. 2008a. An effective aquaculture extension system from farmer's perspective: The case of selected villages in Morogoro and Dar es Salaam regions, Tanzania. *Tanzania Journal of Development Studies* 8(1): 23–35.

Wetengere, K. 2008b. Economic factors critical to adoption of fish farming technology. A case of selected villages in Morogoro and Dar es Salaam regions. Unpublished Ph.D. thesis, Open University of Tanzania, Dar es Salaam.

Wetengere, K. 2009. Socio-economic factors critical for adoption of fish farming technology: The case of selected villages in Eastern Tanzania. *International Journal of Fisheries and Aquaculture* 1(3): 28–39.

Wetengere, K. 2010a. Determinants of adoption of a recommended package of fish farming technology: The case of selected villages in Eastern Tanzania. *Advance Journal of Food Science and Technology* 2(1): 55–62.

Wetengere, K. 2010b. Socio-economic factors critical for intensification of fish farming technology. A case of selected villages in Morogoro and Dar es Salaam regions, Tanzania. *Aquaculture International* 19(1): 33–49. doi:10.1007/s10499-010-9339-2.

Wetengere, K. 2010c. Realizing farmer's objectives—vital to adoption process of fish farming technology. The case of selected villages in Eastern Tanzania. *Advance Journal of Food Science and Technology* 2(2): 115–124.

Wetengere, K. 2010d. Constraints to marketing of farmed fish in rural area. The case of selected villages in Morogoro region, Tanzania. *Aquaculture Economics and Management* 15:130–152.

World Bank. 1989. *World development report*. Washington, DC: World Bank.

The Traditional Whedo Aquaculture System in Northern Benin

MELANIE E. HAUBER[1], DAVID BIERBACH[2], and KARL EDUARD LINSENMAIR[1]

[1] *Department of Tropical Biology and Animal Ecology, University of Würzburg, Würzburg, Germany*
[2] *Department of Ecology and Evolution, J.W. Goethe University Frankfurt, Frankfurt am Main, Germany*

The traditional whedo system involves the digging of large ditch-like ponds that fill seasonally with flood water, trapping fish which are subsequently harvested during the dry season. Although whedos are well described in the South of Benin, basic information necessary for their management in the North is lacking. Since 1998, the commune of Malanville has experienced a kind of 'blue revolution' because up to now, approximately 500 whedos (known locally as tschifi dais) have been constructed covering a total surface area of 9.3 ha, thus reflecting the high acceptance of the practice among the local population. Despite the fast development this system has received little attention from national development planners and international agencies, but in view of the disastrous state of the river Niger's fishery, sustainable aquaculture, fully relying on local species should be supported to release pressure on natural fish stocks. Instead of focussing all efforts on the introduction of modern aquaculture, it is important to enhance the existing system because of the knowledge and acceptance by the local population.

Our sincere thanks go to Assongba Danhin Yao Norbert (Fishery Association of Malanville) for his own initiative in training and assisting the fish farmers and for his technical support and experience-based advice and help. Moreover we express our gratitude to Amidou Nourou-Dine, Bossou Arouna, Bako Mourei, and Nsidou for their assistance in guiding the research team to the different sites. We thank Ms. Monique Sabo Gado for the translation in the course of the interviews. This project was funded by the BIOTA-West project of the BMBF Germany Nr.: 813393-6 01 LC 0017A2.

This paper characterizes the current management and the technical as well as economical features of the tschifi dais in the commune of Malanville, northern Benin.

INTRODUCTION

Malanville (North-East-Benin) is located 1 130 km upstream of the estuary of the River Niger (Moritz *et al.*, 2006). With a length of 4183 km, the Niger is the largest river in West-Africa. In Benin it covers an area of 274 km² at peak flood and forms along 140 km the north-eastern frontier (Welcomme, 1985). The Sota is a tributary with a length of 250 km flowing into the Niger at Malanville (Van den Bossche and Bernacsek, 1990). Almost two million people live in the Niger Basin in Benin (Andersen *et al.*, 2005) and these rivers play an important role in the life of this large and densely populated region by directly contributing to agriculture, fisheries, hunting, grazing and water resources.

Regional fishing is an important traditional activity (Prodecom, 2006); however, local fishermen, fishmongers and consumers complain about the decline of local fish diversity, its quantity and its diminishing size. Reasons for this phenomenon are inter alia an increasing fishery pressure in concert with the utilization of disastrous fishing methods, caused by the increasing demand of a steeply growing population as well as the destruction of spawning areas as a consequence of lacking integrated approaches to floodplain management. Major reasons for the reduction of floodplain area are dam constructions and the plain's deformation for agricultural purposes and habitation. Consequently, fishermen but also farmers suffering from increasing crop failures as effect of the climate change are forced to diversify their sources of income. Therefore, they started to cultivate fish in whedos, *i.e.* in ponds dug in the floodplain of the main rivers. Fish migrating into the inundation zone during flooding become trapped when the floodplain drains during the dry season. Some feeding and management occur and high yields from these whedos have been reported (Welcomme, 1975a). Although, this is a traditional practice of fish rearing in the Ouémé River Valley in southern Benin, it was not known in Malanville before 1998 and the whedos, in the North called 'Tschifi dai' ('fish hole' in the local Dendi language), differ significantly from those in the South of Benin (see Laleyè *et al.*, 2007) especially with regard to their history of development, management strategies, dimensions, climatic conditions, physico-chemical water parameters and fish species diversity.

Since their introduction in 1998, growing fish in tschifi dais has become an important activity. In 2007, we identified 464 tschifi dais covering a surface area of approximately 9.3 hectare. Nowadays, the system is well integrated into the lives of many farmers and their families and plays an important role in supplementing earnings and protein supply especially for the extremely poor.

In view of their positive role in increasing fish production, and the enhancement of livelihoods, a better knowledge of the present state of whedo management is necessary.

MATERIALS AND METHODS

This analysis is based on personal observation, participation in management and interviews with fish farmers, fishmongers and consumers over the period 2007 to 2009. We visited 12 villages in four districts of northern Benin (Garou, Toumboutou, Madekali and Malanville) known to practice the tschifi dai-system and collected information from 49 fish farming groups (all of the tschifi dais we studied were managed by some kind of collective either community or family based).

In all tschifi dais, fish were sampled using a small seine (2mx1m, mesh size 5mm). At final harvest, fish were caught with a big seine (20mx1m, mesh size 5mm), sorted, counted and weighted according to species and total productivity was recorded. Fish were identified according to Paugy *et al*. (2003a, b), Lévêque *et al*. (1990, 1992), Trewavas (1983) and Moritz *et al*. (2006). The positions of the Tschifi dais shown were recorded with the global position system Garmin Etrex Legend.

GENERAL KNOWLEDGE OF THE TSCHIFI DAIS

History and Development

The whedo system, locally named tschifi dai-system, was introduced to the community of Malanville by the CeCCPA (Centre Communale pour le Promotion Agricole) in 1998 which furthermore provided technical support and training for several farmers and fishermen. Thus, the system is no traditional activity although fishing in floodplain depressions and marshes is common in the region and some individuals had already started to practice fish faming in small depressions left in the floodplain after the excavation of clay for brick construction in 1989.

Social Structure

Generally, fish farmers can be considered as agro-fishermen; 91 % stated that their major income comes from agriculture with fishing as a sideline. Only

4.5 % characterize themselves as fulltime fishermen. Nowadays, fishing as main profession is hardly sufficient to support livelihoods. Indicated reasons are the plummeting fish catch, the reduced size of the fish and the increase in labour to catch one unit of fish. Consequently, people are forced to diversify their sources of income, thus becoming agro-fishermen, animal breeders, providers of services, etc.

People usually unite to form associations of 22 members on average (min. 3; max. 80 members) possessing an average of three tschifi dais (min. 1; max. 8). Fish farming seems to be a man's business and generally women are not integrated in the fishing activity but we met one association consisting exclusively of women. Nearly all (90 %) of the groups are hierarchically organized possessing a president, a secretary and a treasurer; each of them responsible for specific tasks. Other positions are organizer, guard and salesman.

Regarding the land use rights, 78 % of the groups are led by landowners who inherited their land as previously noted by Dossou (2008). Most of the rest of the land used for tschifi dais is on loan from the community, meaning from the village chief. A big part of the land used for Whedo construction, mainly marshy lowland, is not suitable for cultivation because of intense flooding during a long period of the year. In return for the access to the land, the fish farmers acknowledge the chief with gifts in the form of fish.

Structural Diversity

Tschifi dais (TDs) can be distinguished according to their location (Fig. 1) and hydrology:

- Category 1: TDs flooded directly by the overspill of rivers;
- Category 2: TDs flooded indirectly through overflowing irrigation channels and rice fields;
- Category 3: TDs only filled by ground water and local rain.

Category 1: The majority of the TDs are dug directly into the fringing floodplain and become connected with the river through heavy local rainfall and drainage of the tributaries. This inundation, 'hari kouarè', has its peak in August/September. Subsequently, a second flood, 'hari bi', occurs in the dry season (November/December) through the overspill of the river Niger as a result of the heavy rainfalls in the highlands of Guinea. However, not all TDs become flooded twice. Ponds situated along the river Sota or in the borderland of the floodplain only become flooded a second time in years of extraordinary intensity of the 'hari bi'.

Category 2: In the vicinity of Malanville, TDs are situated along the edge of rice cultivation. Since this area is protected against the flood by a huge dam these TDs are inundated by overflowing irrigation channels and rice fields, connecting them only indirectly with the river.

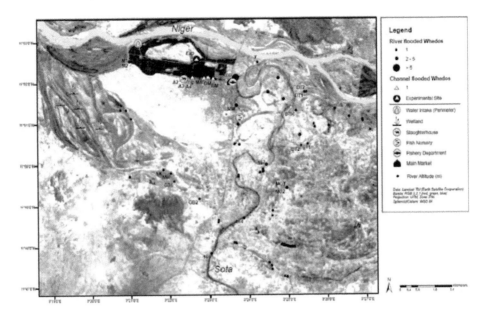

FIGURE 1 Map of Malanville and surrounding area with the studied tschifi dais. The black area along the Niger River represents the area arable for rice cultivation.

Category 3: These fish holes cannot be called tschifi dais in the narrow sense because they are only connected with the rivers in years of unexpectedly heavy inundations. Usually these fish holes are replenished by local rain and ground-water. In the majority they are the result of bad planning and a big part of them dry out with the advancing the dry season.

The TDs have a median size of 200 m² (mode: 200m², n = 74, min. 35.6 m²; max. 7170 m²) with varying depth according to the season. During the rainy season in 2008 the water level averaged 92 cm, and dropped to 52 cm during the dry season, excepting those that dried out completely. Secchi Disk visibility varies from a mean of 12.8 cm during the dry season to 29.5 cm during the rainy season. Most tschifi dais are covered with a dense mat of vegetation mainly consisting of free floating plants such as *Pistia stratiotes* and *Eichhornia crassipes,* but also *Neptunia oleracea, Ludwigia* spp. and *Poacea*-species such as *Echinochloa colona* and *Paspalum scrobiculatum* (Fig. 2).

MANAGEMENT STRATEGIES

Maintenance

Farmers largely neglect their tschifi dais and maintenance mainly refers to the removal of macrophytes at the end of the season to facilitate fish capture. At final harvest, some groups take advantage of the low water level and

FIGURE 2 Removal of macrophytes to facilitate harvest.

remove the bottom mud to reinforce the banks. In contrast to the practice in the Ouémé Delta, gardening or cultivation on the banks is not practiced since fish holes are usually too far away from the homestead and thus difficult to keep under surveillance. Theft is especially a problem in the villages where some tschifi dais were completely emptied. Consequently, some groups place branches of trees in the pond to prevent the use of fishing nets. Additionally, the branches also impede animals, especially cattle, to enter the pond.

Feeding Regime

Approximately 70% of the groups have started to supply supplementary feed during the dry season when the TDs are accessible; however this feeding is very irregular. Although the majority of groups maintained to feed between once per day and once per week, we could not proof this assertion by our observations. It is more common to feed at the beginning of the flood to attract fish or cause them to stay in the TD. Rice bran is by far the most used supplemental feed since it is free and easily available from the numerous local mills. Bran from millet, maize and sorghum are also used but to a lesser extent because they are also fed to ruminants and have to be purchased on the market if not produced on-farm. A few groups also use kitchen-waste or by-products from the slaughterhouse *e.g.* blood and bone meal. Feed is usually mixed with cold or hot water and offered in the form of palm-sized dough-balls. Some groups add clay to guarantee that the dough-balls sink to the bottom. The vast majority of fish growth is due to the natural food supply.

Stocking Regime

The kind of stocking depends primarily on the location of the respective TDs. Category 3 has to be stocked artificially, but nearly 56 % of the other groups also stated that they stock their ponds in addition to the natural fish supply. Most commonly stocked species are *Clarias* sp., *Tilapia* sp., and *Heterotis niloticus*. The majority purchase their fingerlings from local fishermen, though it is not uncommon for tschifi dai operators to collect fish themselves, from, for example, rice fields. Restocking also takes place after the annual harvest with some groups putting fingerlings back to grow some more. However, this stocking is more or less haphazard, with no consistent strategy or knowledge of appropriate stocking. Basically, farmers just take whatever they can get at any time. Our observation showed that artificial stocking of other species than *Clarias* did under the given circumstances not significantly increase the overall biomass at final harvest. Nine tschifi dais were additionally stocked with *Tilapia* spp. (mostly *O. niloticus* and *S. galilaeus*) of different size and density without removing *Clarias* with the result that there was little or no survival to harvest. Most of the harvest is thus mainly a consequence of the degree of natural stocking. Standing stock at harvest was 1.2 kg/m^2 on average.

EXPLOITATION

Period and Frequency

The period of exploitation depends on the extent of the annual floods and the location of the tschifi dai within the floodplain, but the main season is from February to April (Fig. 3), *i.e.* just prior to the onset of the rainy season. In 2007 and 2009 most of the TDs were harvested between March and April because of the relatively minor flood in the previous years, whereas in 2008 the majority was harvested in May reflecting the high water level due to the strong inundation in 2007. Although season and water level are the key indicators of harvest time, farmers also consider the market price of fish. In general, fish prices increase significantly at the end of the dry season because of low landings from the river fishery. Farmers consequently attempt to wait as long as possible into the dry season to maximize the value of their harvest on the local market.

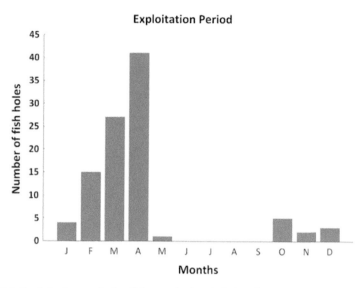

FIGURE 3 Exploitation periods of the studied tschifi dais from 2007 to 2009 (n = 111).

The majority of the TDs (74 %) are harvested in one day by dragging a seine net several times along the length of the pond (Fig. 4). Bigger or highly productive tschifi dais may be harvested several times at weekly intervals.

Gears and Fishing Methods

At harvest, 75 % of the groups interviewed use one or several motor pumps to lower the water level of the TD. The majority of the farmer posses motor pumps for the cultivation of rice and vegetables during the dry season. After lowering the water level, farmers use a big seine (approximately 20m long x 1m deep, mesh size 5mm). Some groups don't remove the vegetation cover completely but leave some patches of floating macrophytes to serve as shelter for the fish and thus preventing them from hiding in the mud. When netting is completed, some groups continue to search for fish hidden in the mud. For this purpose they use harpoons, wooden baskets, sieves or simply their hands. Children are also involved in the harvest stealing small and economically unimportant fish and hiding them in their pockets. Though the owners are aware of this, it is generally considered as compensation for their work. Some groups hire labour for the harvest which is usually paid with fish.

Harvested fish are either kept in large metal bowls sometimes filled with water, woody baskets or in small water holding depressions dug next to the pond. The respective method depends on the locality. In the city bowls are most common whereas in the villages farmers usually dig small pools. The

FIGURE 4 Tschifi dai at final harvest at the end of the dry season. Showed is the dragging of the net and the remaining patches of macrophytes.

main reason why farmers keep the living fish in water is to prevent a loss of body weight and value.

CATCH AND DISPOSITION OF HARVESTED FISH

Diversity and Adaptation of Fish

Prior to the flood, the species diversity of the TDs is highly reduced compared to the fish species recorded after the flood (Hauber *et al.*, in press). With regard to two selected TDs, we observed a species reduction of 64 % and 75 % (from 11 to 4 and from 24 to 6 species, respectively), respectively by comparing the species richness shortly after inundation with that at the end of the dry season. Table 1 lists the species captured at the end of the dry season with their local *Dendi* names.

The majority of the TDs showed low oxygen levels (often 2 mg/l or less) and very high salt concentrations (>4000 μS/cm), which, in addition to high evaporation rates during the dry season, were consequences of the large amounts of decaying organic matter from autochthonous and allochthonous sources. As a result of pollution from human settlements,

TABLE 1 Fish Species Detected at the End of the Dry Season and Their Local Names

Species	Local name (*Dendi*)
Protopteridae	
Protopterus annectens	Siyibi
Polypteridae	
Polypterus senegalus	Gondo-Kououga
Osteoglossidae	
Heterotis niloticus	Koualah
Mormyridae	
Brevimyrus niger	Wassi
Claroteidae	
Auchenoglanis occidentalis	Koutoukou tchiré, Bouro
Clariidae	
Clarias gariepinus	Dessebi, Dessi
Heterobranchus longifilis	Dessi tchiré
Cyprinodontidae	
Epiplatys spilargyreius	–
Nothobranchius kiyawensis	–
Channidae	
Parachanna obscura	Corombou
Cichlidae	
Hemichromis cf. *letourneauxi*	Koula-Koula
Oreochromis niloticus	Fotoforoh-Bi, Kossia-Bi
Sarotherodon galilaeus	Kossia-Koare
Anabantidae	
Ctenopoma sp.	–
Mochokidae	
Synodontis schall	Djidjiri, Koutoukou
TOTAL	15

total anoxia not uncommonly causes catastrophic die-offs. Thus it came as no great surprise that all species detected at the end of the dry season are known for their ability to survive low oxygen levels. They are either using atmospheric air, as *e.g. H. niloticus* (Lüling, 1977); *P. senegalus, P. annectens, Clarias* spp. (Welcome, 1979); *Heterobranchus longifilis* (Holden and Reed, 1978); *Brevimyrus niger* (Moritz & Linsenmair, 2007), or the oxygen-rich surface layer of the water, as *e.g. O. niloticus, S. galilaeus, Synodontis* spp. (Bénech & Lek, 1981), killifishes (Lewis, 1970; Kramer & McClure, 1982). Furthermore, some species, such as cichlids and catfishes, are known for their ability to withstand high salt concentrations (Eddy *et al.*, 1980; Foskett *et al.*, 1981; Whitfield & Blabber, 1976).

Total and Species-Specific Annual Biomass Harvested

Changes in species composition and richness in the TDs post-flood compared to their status at the end of the dry season is most obvious at final harvest. Conspicuous is the dominance of only one genus (*Clarias* spp.) in almost all investigated tschifi dais by the end of the dry season. In the 22

TABLE 2 Fish Species Recorded at Final Exploitation and Their Percentage on the Total Biomass Harvested

Species Group	Clarias sp.	Heterotis niloticus	Tilapia spp.	Polypterus senegalus	Protopterus annectens	Heterobranchus longifilis	Others*
2008							
DamatabiNsidou	99,6	0,0	0,0	0,4	0,0	0,0	0,0
Noma Idé I	100,0	0,0	0,0	0,0	0,0	0,0	0,0
Noma Idé II	97,4	0,0	1,7	0,0	0,0	0,0	0,9
Gounai Windi	93,9	6,1	0,0	0,0	0,0	0,0	0,0
Lakalikanei I	92,0	0,0	6,7	0,0	0,0	1,4	0,0
Lakalikanei II	66,3	0,0	12,6	0,0	0,0	0,0	21,2
Dagara Mama	87,5	11,8	0,7	0,0	0,0	0,0	0,0
Macheresse	38,6	27,1	29,8	1,4	3,1	0,0	0,0
2009							
Botcho Manou	99,0	0,0	0,4	0,6	0,0	0,0	0,0
DamatabiNsidou	90,8	7,6	0,0	0,6	0,1	0,0	0,9
Noma Idé Village	12,8	24,1	0,0	0,0	0,0	63,1	0,0
Noma Idé I	100,0	0,0	0,0	0,0	0,0	0,0	0,0
Noma Idé II	99,6	0,0	0,0	0,3	0,0	0,0	0,1
Goro Bani I	89,6	0,0	9,0	1,4	0,0	0,0	0,0
Macheresse	24,8	18,0	56,8	0,0	0,4	0,0	0,0
Mamamassou	86,5	0,0	13,5	0,0	0,0	0,0	0,0
Mourou Windi	98,9	0,0	1,1	0,0	0,0	0,0	0,0
Heoufounin 3	100,0	0,0	0,0	0,0	0,0	0,0	0,0
Heoufounin 4	96,4	0,0	1,3	0,0	2,3	0,0	0,0
Heoufounin 5	96,6	0,0	0,0	0,0	3,4	0,0	0,0
Heoufounin 6	95,8	0,0	4,2	0,0	0,0	0,0	0,0
Heoufounin 7	25,6	0,0	61,7	0,0	9,7	0,0	3,1

*= *Brevimyrus niger*, *Ctenopoma* sp., *Auchenoglanis occidentalis*, etc.

harvests in which we participated from 2008 to 2009, *Clarias* accounted for 81.4% on average of the total biomass, whereas in more than 50 % of the TDs *Clarias* even constituted 95% of the total biomass. Table 2 gives an overview of the different species caught and their percentage of the total biomass at final harvest. An average of 93 % of the total biomass of *Clarias* caught consists of small and medium sized fish (min. 22 g; max. 450 g), whereas only 7.6 % of the harvest consists of fish with an average weight of 900 grams (min. 500 g; max. 1600 g). The majority (72 %) consists of *Clarias* with an average weight of only 38 grams (min. 22 g; max. 60 g).

By converting the biomass harvested on a hectare basis separately for each of the TDs, the annual biomass averaged 11.8 tons/ha in 2008 and 2009 (see Tab. 3). However, as previously mentioned, biomass is highly dependent on the flood regime of the previous year (Welcomme and De Mérona, 1988), thus the influence of the extensive flooding in 2007 can be seen in the average yields of 17 tons/ha in 2008 and 8.6 tons/ha in 2009, following a relatively dry 2008. In years of weak flooding, the biomass yield of the TDs consists mostly of fish left from the previous year.

TABLE 3 Biomass Output of the Different Whedos from 2008 to 2009

Group	Total weight (kg)	Surface area (m^2)	Output (kg/ha)
2008			
Lakalikanei I	562,9	271,8	20712,1
Lakalikanei II	8,8	36,6	2390,7
Macheresse	513,6	7170,0	716,2
Dagara Mama	6,8	142,7	476,6
Eloa Madekali	235,8	114,0	20684,2
Gounai Windi	164,0	108,0	15185,2
Bobosotjiré	455,0	163,4	27845,8
DamatabiNsidou	449,3	61,0	73655,7
Noma Idé 1	88,5	207,0	4275,4
Noma Idé 2	84,7	239,7	3533,6
2009			
Heoufounin 3	3,3	96,2	337,8
Heoufounin 4	79,2	247,0	3207,5
Heoufounin 5	37,1	46,5	7978,5
Heoufounin 6	77,2	314,3	2455,8
Heoufounin 7	11,4	35,6	3184,6
Lakalikanei I	218,0	271,8	8022,1
MamaMassou	137,1	60,7	22570,8
Macheresse	413,1	7170,0	576,1
Botchou Manu	39,0	702,0	555,6
Bobosotjiré	46,0	200,0	2300,0
DamatabiNsidou	172,0	61,0	28188,5
Noma Idé Kotchi	71,5	275,5	2595,3
Noma Idé 1	108,0	418,5	2580,6
Noma Idé 2	99,4	480,0	2069,8
Goro Bani	486,9	170,0	28641,2
Mourou Windi	359,1	160,0	22445,0
Min. biomass output (kg/ha)			337,8
Max. biomass output (kg/ha)			73655,7
Average biomass (kg/ha)			11814,8

When relating the total biomass harvested onto the total water surface of the exploited TDs, annual average yield was 3 t/ha in 2008 and 2.1 t/ha in 2009. Unfortunately, there are no data available on the biomass production of the TDs in Malanville but yields of the 'Whedos' in the Ouémé-Delta ranged from 2.06 tons/ha in 1955 to 1.31 tons/ha in 2001 (Lalèyè *et al.*, 2007; Welcomme, 1971), although it is not clear if the author refers to single TDs or to the total water surface area.

According to case studies of the 'Ebe'-fishery in Ghana, similar whedos/fish holes yield between 13.3 and 26.7 t/ha water surface annually (gtz, 2002).

Although this sounds inconceivable, it should be borne in mind that with the retreat of the flood a large percentage of fish dwelling in the floodplain get trapped within these depressions. Lowe-McConnell (1964) found 870 fish belonging to 36 species in a floodplain pool of only 19 m^3. The fact that the density of whedos installed in the Ouémé Delta is far higher than in the North might explain their lower annual yields.

However, the high differences of output within the TDs in the north are probably a result of their location within the floodplain, the feeding regime and the number of TDs in their close vicinity since our data do not proof any relation between the biomass harvested and the dimension or the age of the TDs.

THE VALUE CHAIN

While fish farming is almost exclusively a males' business, women are integrated within this sector in marketing, processing and distribution of the fish. The majority of the fishmongers (79 %) are relatives of the fish farmers and are directly involved in determination of the harvest date. Since women are usually united in professional associations they also get informed by their association president. Mongers arrive at the site and purchase the fish directly after the harvest. Usually demand is greater than the harvested supply and the waiting fishmongers will evenly share the available fish. The fish are sold in locally accepted scale units consisting of two different sizes of metal basins, the smaller and more common of which, holding an average of 21 kg is known as Weguisé. The price of one basin varies according to the size of the fish. Small fish (average weight 40 g) cost 431 FCFA (0.66 €) per kg while medium sized fish (average 350 g) sell for 536 FCFA (0.82 €) per kg. Basins filled with different size classes of fish generate about 513 FCFA (0.78 €) per kg. Bigger fish are mostly sold individually. In comparison, market prices of fish from the river fishery are generally higher, with one kilogram of fresh *Clarias* selling for 879 FCFA (1.34 €).

After purchasing the fresh fish, mostly *Clarias*, the mongers usually process the fish before reselling, though some are transported alive in water for sale at their homestead. Processing depends on the species; e.g. *Tilapia* spp. is usually fried whereas the majority of *Clarias* sp. is smoked coiled up and gored on a wooden stick. Smoked fish is not only sold on the local markets but also sold on other major markets e.g. in Kandi and exported to Nigeria. One kilogram of smoked Clarias is sold for 2,200 FCFA (3.35 €) on average, while fried Tilapia is offered for 4,245 FCFA (6.48 €). Some women prepare the fish as meals and sell them for consumption on the street.

Consumer preferences differ according to region. The indigenous people around Malanville prefer *Clarias* because it can be easily preserved by smoking. People immigrated from the south prefer *Tilapia* spp. Bigger fish are highly appreciated but most consumers cannot afford the higher prices, so larger specimens are cut up and sold per piece.

INCOME

Income from fish farming can be divided into cash and food. According to the farmers, 11% sell none of the fish but share them between the group members for home consumption and give some to the village chief as a gift. However, the majority of farmers are cultivating fish to earn some additional income and 65% of the farmers said they sell more than 80 % of the harvest. Some 62% of the farmers save a part of the money by depositing it at the CLCAM (Caisse Local de Crédit Agricole Mutuel) or at the treasurer. The rest of the money is divided between the members of the whedo association. Farmers confirm that earnings derived from the TDs are of great importance since they fall prior to the onset of the new agricultural season when farmers need the money for purchasing seeds, agricultural tools etc. Weddings, baptisms and other events as well as medical treatments and emergency cases are also paid for with the income of aquaculture activities.

The income depends on the size as well as the stocking density of the respective TD. Realized gross income ranges between 68,758 (104.81 €) and 327,000 FCFA (498.5 €) not counting home consumption and donations. The total surface area of the TDs with available economic data was 18,091 m² generating an income of 1,990,482 FCFA (3,034.3 €), leading to 1,099,953 FCFA /ha (1,676.80 €). If we convert the outcome to the total 9.3 ha of surface water area under production, the annual gross income of the commune attains 10,207,562 FCFA (15,560 €). This is similar to the findings of Imorou Toko (2007) who reported that fishing in 1,050 whedos can yield more than 25 million FCFA (39,117 €). However, these are only gross incomes since we did not consider any costs for construction or inputs such as feed or fingerlings. Usually construction is done directly by the group members and running costs are rather low since feeding is negligible and usually done with products free of charge.

MAJOR CONSTRAINTS

Despite the fast development of the TDs there exist some drawbacks hindering the improvement of the current system. In Madekali, fish farmers are in conflict with local fishermen since their TDs are situated in a small branch of the River Niger and the fishermen accuse them of being somehow responsible for reduced fish landings. In Malanville, three fish farmer associations are in conflict with the neighbouring school that raised a claim on the land although the farmers possess deeds of ownership. However, the most important drawback is theft. These detrimental impacts discourage the owners and deter them from maintaining and feeding their TDs. It is also important to mention that there are also internal problems within groups and a number of associations have already disbanded because of the misbehaviour of the some members affecting the whole alliance.

CONCLUSIONS

Because of rapid demographic growth and the overexploitation of the rivers, it is important to look for alternatives to increase fish supply in a sustainable way. In Malanville, the consumption of chilled fish increased from 16,630 kg in 1999 to 33,000 kg in 2007 (unpublished data of the local fishery department). Fish farming in TDs is an attractive system for the rural population because of existing knowledge of post-flood wetland fisheries *e.g.* fishing in marshes and natural depressions in the floodplain, as well as the low investment needed for its installation. Natural stocking also reduces the cost of purchasing fingerlings and solves other procurement problems.

However, the system is still in its infancy. The fact that 72% of the total catch consists of *Clarias* with an average weight of 40 grams, highlights the need to increase the knowledge among the farmers that supplementary feed will enhance their yields. Farmers should also concentrate their efforts on improving water quality *e.g.* by avoiding sites that are highly polluted. Moreover, it is essential to increase farmers' knowledge on appropriate stocking such as species, size, density and ratio of fish to be stocked and other methods to reduce wasted capital and manpower.

If only 1% of Africa's 12 million hectares of floodplains would be developed as whedos with an output of 1 t/ha/y, the potential yield would be as much as 120,000 tons per year (Balarin, 1988). But despite this high potential, further studies are needed to clarify whether increasing whedo construction might transform the nature of the floodplain in a way which will negatively affect its fish community and biodiversity.

According to Welcomme (1975b), there is no risk of overfishing since the provision of whedos retains fish that would otherwise escape to the river where there would be inadequate living space for them or that would die through eventual desiccation of the pools and as well from facilitated predation through predators such as birds and piscivorous fish due to the very restricted escape possibilities. Surely, the provision of TDs will increase the area of the plain that holds water during the dry season, thus extending the habitat for air breathing and floodplain dwelling species. Consequently, production from the floodplain could be improved. However, to prevent recruitment overfishing it should be secured that the whedos are not desiccating and that farmers use nets with an appropriate mesh size (> 20mm) to guarantee the escape of fingerlings.

It is also not clear if deep ponds might affect riverine species *e.g.* by causing them to stay longer on the floodplain resulting in their premature death because of decreasing water quality. But a high density of the TDs might also have an indirect impact by creating a deranged balance of the species assemblage as a result of an increased appearance of adapted species. Therefore, a drastic extension of the surface area of whedos should be avoided or at least the installation of protected spawning areas should be guaranteed.

ACKNOWLEDGEMENTS

Our sincere thanks go to Assongba Danhin Yao Norbert (Fishery association of Malanville) for his own initiative in training and assisting the fish farmers and for his technical support and experience-based advice and help. Moreover we express our gratitude to Amidou Nourou-Dine, Bossou Arouna, Bako Mourei and Nsidou for their assistance in guiding the research team to the different sites. Ms. Monique Sabo Gado we thank for the translation in the course of the interviews. This project was funded by the BIOTA-West project of the BMBF Germany Nr.: 813393-6 01 LC 0017A2.

LITERATURE

Andersen, I., O. Dione, M. Jarosewich-Holder, and J.-C. Olivry. 2006. The Niger River basin: A vision for sustainable management. K.G. Golitzen, ed. The World Bank. Washington, DC. 144p.

Balarin, J.D. 1988. Aquaculture developing planning: The logistics of fish farm project appraisal. Pages 92–106 in H.R. King, and K.H. Ibrahim, eds. Village level aquaculture development. Proc. of Commonwealth consultative workshop, Freetown, Sierra Leone. Commonwealth Secretariat Publ.

Bénech, V., and S. Lek. 1981. Résistance à l'hypoxie et observations écologiques pour seize espèces des poissons du Tchad. *Rev. Hydrobiol. Trop.* 14: 153–168.

Dossou, S. 2008. Studies on the prospects and constraints of the development of aquaculture in the commune of Malanville in Benin. DEA Thesis. University of Abomey-Calavi, Benin. Faculty of Agronomy. 83p.

Eddy, E.S., O.S. Bamford, and G.M.O. Maloiy. 1980. Sodium and Chloride balance in the African Catfish *Clarias mossambicus*. *Comp. Biochem. Physiol.* 66, 637–664.

Foskett, J. K., D. Craig, D. Logsdon, T. Turner, T.E. Machen, and H.A. Bern. 1981. Differentiation of the Chloride extension mechanism during seawater adaptation of a teleost fish, the Cichlid *Sarotherodon mossambicus*. *J. Exp. Biol.* 93: 209–224.

Hauber, M., D. Bierbach, and K.E. Linsenmair, in press. A description of teleost fish diversity in floodplain pools ('Whedos') and the Middle-Niger at Malanville (North-eastern Benin). Journal of Applied Ichthyology.

Holden, M.J., and W. Reed. 1978. West African freshwater fish (West African nature handbooks). Longman Group, London, England. 68p.

Gtz. 2002. Back to Basics. Traditional Inland Fisheries Management and Enhancement Systems in Sub-Saharan Africa and their Potential for Development. Universum Verlagsanstalt, Wiesbaden, Germany. 203p.

Imorou Toko, I. 2007. Amélioration de la production halieutique des trous traditionnels à poissons (Whedos) du delta de l'Ouémé (sud Bénin) par la promotion de l'élevage des poissons-chats *Clarias gariepinus* et *Heterobranchus longifilis*. Dissertation of the Faculties of Sciences, Institute of Biology; University Notre-Dame de la Paix, Namur, Belgium. 186p.

Lalèyè, P., D. Akélé, and J.-C. Philippart, 2007. La pêche traditionnelle dans les plaines inondables du fleuve Ouémé au Bénin. *Cahiers d'Ethologie* 22 (2): 25–38.

Lévêque, C., D. Paugy, and G.G. Teugels. 1990. Faune des poissons d'eaux douces et saumâtres de l'Afrique de l'Ouest. Volume 1. Collection Faune et Flore tropicales 28. IRD, Paris.

Lévêque, C., D. Paugy, and G.G. Teugels. 1992. Faune des poissons d'eaux douces et saumâtres de l'Afrique de l'Ouest. Volume 2. Collection Faune et Flore tropicales 28, IRD, Paris.

Lowe-McConnell, R.H. 1964. The fishes of the Rupununi savanna district of British Guiana, South America, Part 1. Ecological groupings of fish species and affects of the seasonal cycle on fish. *J. Limn. Soc. (Zool.)* 45: 103–144.

Lüling, K.-H. 1977. Die Knochenzünglerfische. A. Ziemsen Verlag, Wittenberg.

Moritz, T., P. Lalèyè, G. Koba, and K.E. Linsenmair. 2006. An annotated list of fish from the River Niger at Malanville, Benin, with notes on the local fisheries. *Verhandlung der Gesellschaft für Ichthyologie* 5: 95–110.

Moritz, T., and K.E. Linsenmair. 2007. The air-breathing behaviour of *Brevimyrus niger* (Osteoglossomorpha, Mormyridae). *J. Fish. Biol.* 71: 279–283.

Paugy, D., C. Lévêque, and G.G. Teugels (eds). 2003a: Faune des poissons d'eaux douces et saumâtres de l'Afrique de l'Ouest. Tome 1. Collection Faune et Flore tropicales 40. IRD, Paris.

Paugy, D., C. Lévêque, and G.G. Teugels (eds). 2003b: Faune des poissons d'eaux douces et saumâtres de l'Afrique de l'Ouest. Tome 1. Collection Faune et Flore tropicales 40. IRD, Paris.

Prodecom, 2006. Schéma directeur d'aménagement de la commune de Malanville: analyse et diagnostic. SERHAUS-SA, Cotonou-Benin. 23p.

Trewavas, E. 1983. Tilapiine fishes of the genera *Sarotherodon*, *Oreochromis* and *Danakilia*. British Museum (Natural History), London.

Welcomme, R.L. 1971. Evaluation de la pêche intérieure au Dahomey, son état actuel et ses possibilités. FAO AT 2938, Rome, 95p.

Welcomme, R.L. 1975a. The fishery ecology of African floodplains. CIFA Technical Paper 3. FAO, Rome.

Welcomme, R.L. 1975b. Extensive Aquaculture Practices on African Floodplains. *In* CIFA Technical Paper (4) Supplement 1 to the Report of the Symposium on Aquaculture in Africa, Accra, Ghana, 30 September – 2 October 1975. Reviews and experience papers. 791p.

Welcomme, R.L. 1979. Fisheries ecology of Floodplain Rivers. London, Longman, 317p.

Welcomme, R. L. 1985. FAO Fisheries Technical Paper 262, River fisheries. FAO, Rome

Welcomme, R.L., and B. de Mérona. 1988. Fish communities of rivers – Peuplements ichtyologiques des rivières. Pages 251–276 *in* C. Lévêque, M. Bruton, and G. Ssentongo, eds. Biologie et écologie des poisons d'eau douce africains. Paris, France: ORSTOM, 216.

Whitfield, A. K., and S.J.M Blaber. 1976. The effects of temperature and salinity on Tilapia rendalli, Boulenger 1896. *J. Fish. Biol*. 9: 99–104.

Van den Bossche, J. P. and G.M. Bernacsek. 1990. Source book for the inland fishery resources of Africa. CIFA Technical Paper Vol. 2 (18/2). Rome.

Production Parameters and Economics of Small-Scale Tilapia Cage Aquaculture in the Volta Lake, Ghana

J. K. OFORI[1], E. K. ABBAN[1], A. Y. KARIKARI[1], and R. E. BRUMMETT[2]

[1] Water Research Institute, P.O. Box M32, Accra, Ghana
[2] WorldFish Center, P.O. Box 2008 (Messa), Yaoundé, Cameroun

To calculate the potential for cage aquaculture to create economic opportunities for small-scale investors on the Volta Lake, Ghana, a local NGO with technical support from the Government of Ghana ran two trials (one of four and one of six units) of small-scale cage aquaculture in the town of Dzemeni. Cages were built locally from available materials at a cost of approximately US$1000 per 48 m^3 cage. An indigenous line of Nile tilapia, Oreochromis niloticus, *was stocked either as mixed sex (first trial) or all-males (second trial) at an average rate of 103 fish/m^3 and grown on locally available pelleted feeds for approximately six months. Total costs averaged US$2038 per six-month production cycle. Gross yield ranged from 232 to 1176 kg/cage, averaging 460 kg/cage (9.6 kg/m^3). Final average weight of mixed sex populations (253.05 ± 47.43g) was significantly less than of all-males (376.7 ± 72.30g). Likewise, percentage of fish over 300 g at harvest was significantly lower in mixed-sex (38.3%) compared to all-male (75.7%) populations. Mortality resulting primarily from poor handling during transport and stocking averaged 70% and was a major determinate of production and profitability. To break even, harvested biomass of fish needed to exceed 15 kg/m^3. At 25 kg/m^3, small-scale cage aquaculture generated a net income of US$717 per cage per six months (ROI = 30.2%) on revenues of US$3,500.*

This research was co-funded by Rural Wealth and the CGIAR Water and Food Challenge Program Project 34: Increasing Fish Production from the Volta Lake. Special thanks to Tropo Farms, Ghana; Lake Harvest, Zimbabwe, and the FISH-Uganda project for sharing production data.

Water quality in the area surrounding the cages was not negatively affected by aquaculture at the scale tested (5 tons of feed per six months).

INTRODUCTION

Along the Volta Lake in Ghana, increasing population, rising unemployment, and declining value of the capture fishery are driving efforts to identify viable investment opportunities that can increase revenues and promote sustainable economic growth (Abban et al. 2006). Cage aquaculture is practiced profitably in many parts of the world, generating jobs and making substantial contributions to fish supply (Hambrey 2006). The FAO (2007) estimates that the Atlantic salmon (*Salmo salar*) cage aquaculture industry alone produces over 1.2 million MT per annum. From interviews with the major producers, current output of cage aquaculture in sub-Saharan Africa, almost all of which is Nile tilapia, *Oreochromis niloticus*, can be estimated at about 5500 MT per annum, mostly from relatively large-scale projects in Ghana, Malawi, Uganda and Zimbabwe.

The Volta Lake is a reservoir on the Volta River created by the construction of the Akosombo Dam, completed in 1964. Temperature, flow rates, and water quality are generally high, while fertility of the water is relatively low, reflecting the lake's oligotrophic status (Abban & Biney 1996). The Government of Ghana has allocated 1 percent of the total surface area (8,700 km^2) of the lake to the development of cage aquaculture by (personal communication, Hon. Gladys Asmah, Ghana Minister of Fisheries, Accra, 30 March 2007). If the yields reported for cage aquaculture elsewhere in Africa—between 50 (personal communication, Patrick Blow, Lake Harvest Aquaculture, 31 October 2006) and 150 kg/m^3/9 months (personal communication, Karen Veverica, Jinja, Uganda, 21 November 2007)—can be replicated in Ghana, production from less than 100 ha of fish cages could just about match the current capture fishery output of about 90,000 MT (Asante 2006).

In Ghana, positive cash flows have been reported for medium-scale production systems that can achieve outputs of approximately 20 tonnes per month (240 TPA). The stocking, feeding, and cage construction technology piloted by these farms have proven to be generally suitable to smaller-scale investors as well and is being widely adopted in Stratum II of the Volta Lake (Figure 1) and between the Kpong and Akosombo Dams on the lower Volta River.

During the period between October 2005 and December 2007, Rural Wealth, a Ghanaian non-governmental organization (NGO), in a technical partnership with the Ghanaian Water Research Institute (WRI) and the

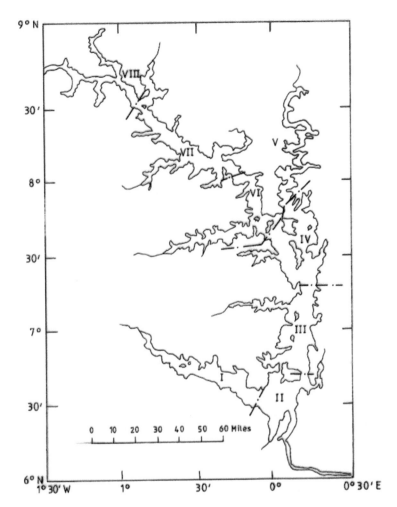

FIGURE 1 The Volta Lake indicating its management strata: I = Afram arm; II = lower main body; III = middle main body; IV = upper main body; V = Oti river arm; VI = lower Volta riverine body; VII = middle Volta riverine body; and VIII = ipper Volta riverine body (Petr & Vanderpuye 1964).

WorldFish Center, led a demonstration of small-scale cage aquaculture to identify key production constraints and technical difficulties, and test production capacity and economic viability over two six-month growout cycles.

MATERIALS & METHODS

The site (N 06° 36.150, E 000° 09.17) was in open water 0.5 km from shore, in the lee of a small-uninhabited island off of the fishing and trading community of Dzemeni, South Dayi District, Volta Region, Ghana, on the SW bank of

FIGURE 2 Small-scale cage culture installation near the fishing community of Dzemeni in the South of Lake Volta, Ghana.

the Volta Lake in Stratum II. For the first of two trials, four cages each of 6 × 4 × 2 m deep (48 m^3) constructed of the typical 15-mm multifilament stretched mesh netting used in the local beach seine fishery attached to a pipe frame supported by floating plastic barrels were anchored in 8–10 m of water (Figure 2). The anchors were comprised of four 0.3 m^3 concrete blocks into each of which a 6.4 mm (1/4″) iron loop was cemented and to which were tied ropes connected to each corner of the floating cage framework from which the cage was suspended. Cages were constructed entirely of locally available materials (Figure 3) at an individual cost of approximately $1000 (Table 1). For the second trial, an additional two cages were installed for a total of six (Table 2).

Mixed-sex fingerlings (first trial) and all-male fingerlings (second trial) derived from a selected line of *Oreochromis niloticus* produced at the Ghanaian Aquaculture Research and Development Centre (ARDEC) in Akosombo, and reported by WRI to grow some 10–15% faster than the local wild stock, were stocked at rates ranging from 3000 up to 9000 fish per cage (63 to 188 fish/m^3), but heavy mortalities incurred as a result of poor transport and handling and resulted in effective stocking rates of between 20 and 100 fish/m^3 with fingerlings of between 13 and 32 g (Table 2). Dead fish found floating in the cages were removed daily and recorded to give an estimate of mortalities.

Fish in cages were fed locally available (GAFCO Inc.) floating extruded pelleted aquafeed containing approximately 30% crude protein. Fish were

FIGURE 3 Pilot small-scale cage aquaculture pilot design tested at Dzemeni, Ghana, from June 2006 through December 2007. Empty 250 l plastic barrels provide floatation to a 6 × 4 × 2 m (48 m³) 15-mm mesh net suspended from a frame of galvanized pipe bolted into welded brackets.

TABLE 1 Construction Costs for a 48 m³ Small-Scale Aquaculture Cage Manufactured from Locally Available Materials in Dzemeni, Ghana (1 USD = 0.92 Ghana ¢)

Item	Description	Quantity	Unit cost	Amount (Ghana ¢)
Galvanized pipe	1.5"–2"	12	18	216
Floats	Plastic barrel (250 l)	8	30	240
Nets	15 mm stretched mesh	40 m	4.375	175
Shackles		16	3	48
Hapa nets	40 m	0.5	100	50
Rope	10 mm	2 coils	20	20
	6 mm	1 coil	5	5
Anchors	0.3 m³	6	5	30
Welding			50	50
Cage cover net	6 m × 5 m	1	12	12
Labour		1	30	30
Total				876

fed at a declining rate of 10% down to 2% of estimated average bodyweight according to the recommendations of Hepher (1988) based on monthly average weight of a sample of at least 50 fish from each cage. The total daily ration was divided among three feedings administered by hand.

After 133 to 153 days, cages were harvested, fish graded according to size class (>300 g, 200–300 g, <200 g), counted, weighed, and sold to local fish traders. Wholesale prices varied according to size class with >300 g fish sold at Ghana Cedis ¢ 3.50 per kg (1 US$ = 1.12 ¢), 250–300 g fish sold at ¢2.80, and <200 g fish sold at ¢1.50 per kg. Production and price parameters were then compiled and tabulated to analyze the economic performance of small-scale cage culture in the lower Volta Lake.

During the first trial, water quality was monitored every two months. In an attempt to estimate the impacts of cage aquaculture on water quality,

TABLE 2A Fish Stocking, Growth, and Harvest Data for the First (Mixed Sex) of Two Pilot Nile Tilapia 48 m³ Cage Aquaculture Trials Conducted at Dzemeni in Stratum II of the Volta Lake, Ghana

	Cage 1	Cage 2	Cage 3	Cage 4
Date Stocked 2006	27/10	08/12	31/10	31/10
No. Stocked	4000	6000	7000	2780
Avg. wt at stocking (g)	13.4 ± 10.33	25.0 ± 4.14	12.9 ± 7.97	31.7 ± 15.02
Avg. wt at harvest (g)	207.5 ± 59.98	277.5 ± 42.36	219.7 ± 88.27	307.5 ± 134.19
Grow-out (days)	153	147	133	152
Biomass GR (kg/day)	2.69	9.05	7.1	4.14
No. Fish at harvest	1946	4639	1079	1647
Survival (%)	48.70	77.3	15.4	59.2
SGR*	1.79	1.64	2.13	1.49
FCR	2.64	2.50	3.51	2.97
Gross yield (kg/cage)	324.7	1175.7	247.1	402.5
Net yield (kg/cage)	270.9	1025.7	221.3	314.3

TABLE 2B Fish Stocking, Growth, and Harvest Data for the Second (All-Male) of Two Pilot Nile Tilapia 48m³ Cage Aquaculture Trials Conducted at Dzemeni in Stratum II of the Volta Lake, Ghana

	Cage 1	Cage 2	Cage 3	Cage 4	Cage 5	Cage 6
Date Stocked 2007	31/1	6/7	7/3	31/08	18/07	7/7
No. Stocked	7500	8200	7500	9000	6000	6000
Avg. wt at stocking (g)	22.92 ± 9.75	22.88 ± 8.75	12.45 ± 4.92	12.45 ± 4.90	20.08 ± 9.32	20.13 ± 9.34
Avg. wt at harvest (g)	369.70 ±155.50	452.2 ±230.32	Net caught on bottom and torn open; Last sample 10/01/08 avg wt 318 g	Net slashed by fishers; last sample 25/10/07 avg wt 86.2 g	Poisoned by fishers; last sample 23/12/07 avg wt 284 g	308.1 ±141.23
Grow-out (days)	169	147				130
No fish at harvest	1480	523				1542
Survival (%)	20	6.4				25.7
SGR*	1.65	2.03				2.10
FCR	2.64	2.5				8.05
Gross yield (kg/cage)	324.7	232				503.4
Net yield (kg/cage)	270.94	44.38				389.68

*Specific growth rate (%/day) = ln average final weight − average initial weight/growth period (days).

sampling stations were located 10 m upstream and 30 m downstream of the cage installation. Temperature and pH were measured in the field with a temperature probe and a HACH EC 20 portable pH meter, respectively. Samples for analysis in the laboratory were collected from near the top of the water column, at mid-cage depth (1 m) and near the bottom in 300 ml plain glass bottles, kept in a cold dark box during transportation, and stored in a refrigerator until the analyses were completed. In the laboratory, electrical conductivity was measured with a Cyberscan 510 meter and turbidity was measured with a HACH 2100 P Turbidimeter. All other analyses followed standard methods (APHA/AWWA/WEF 1998): diazotization for NO_2-N; hydrazine reduction for NO_3-N; direct nesslerization for total ammonia nitrogen; stannous chloride for phosphate; and the Azide modification of Winkler for dissolved oxygen.

RESULTS

Production and growth data from the trials are shown in Table 2 and Figure 4. Two cages were sabotaged by locals and another was damaged when it became fouled with a submerged tree when the water level was low and then ripped open, releasing the fish, when the water level rose again. For those cages that survived the entire trial, gross yield ranged from 232 to 1176 kg/cage/6 months (5–25 kg/m^3), averaging 456 ± 329.5 kg/cage (9.5 kg/m^3). Overall, survival was low in all cages, averaging 29 ± 28.4% among those that were not damaged or robbed, most of which was incurred

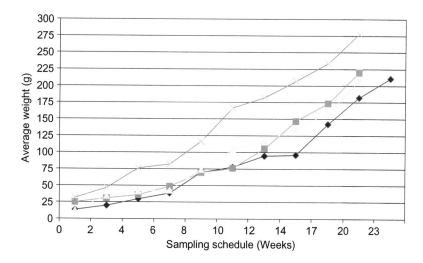

FIGURE 4A Growth pattern of mixed sex *Oreochromis niloticus* in four 48 m^3 cages in the Volta Lake, Ghana; fish stocked at approximately 40 fish/m^3 were fed a commercial diet over a culture period of six months in 2006/2007.

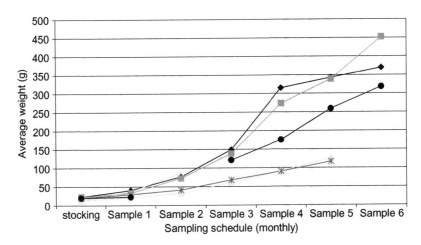

FIGURE 4B Growth pattern of all-male *Oreochromis niloticus* in six 48 m³ cages in the Volta Lake, Ghana; fish stocked at approximately 40 fish/m³ were fed a commercial diet over a culture period of six months in 2007. Three cages were damaged or sabotaged and harvested early.

as a result of poor fish conditioning, handling and transport during stocking. Only about 30% of the mortalities floated up and were counted, at least another 40% sank or went unnoticed.

Only the cage from which more than one tonne of fish (>96 fish weighing 24.5 kg per m³) was harvested made a significant profit (Table 3). Food conversion ratios (FCR) were between 2.5 and 8.1 with an average of 3.54 (Table 2). Feed was the major component of cost, averaging over 50% of the total (Table 4). Fingerling purchase was another major cost, accounting for an average of 27% of the total.

The WRI Akosombo improved strain exhibited an average specific growth rate in cages of 1.83% body weight per day (Table 2), but showed significant variation in final weight at harvest, ranging between 60 and 500 g. Final weight of mixed sex populations averaged 253.05 ± 47.43 g, significantly ($P < 0.05$) less than the 376.7 ± 72.30 g achieved in the all-male populations. Likewise, percentage of fish over 300 g at harvest was

TABLE 3 Economic Analysis of Tilapia Aquaculture Cages Operated for Approximately Six Months in Stratum II of the Volta Lake (Ghana ¢; 1 US$ = 1.12 Ghanac ¢)

	Cage 1	Cage 2	Cage 3	Cage 4	Cage 5	Cage 6	Cage 10
Fixed cost*	109.5	109.5	109.5	109.5	109.5	109.5	109.5
Variable costs	1,760.76	2,700.00	2,458.70	1,468.06	1,598.0	1,867.0	1,687.0
Total cost	1,870.26	2,809.50	2,568.20	1,577.56	1,707.5	1,978.5	1,756.5
Revenue	812.82	3,527.04	741.3	1,207.50	1,136.4	812.00	1,760.1
Net Income	−1,057.44	717.54	−1,826.90	−370.06	−574.1	−895.50	4.50

*For the cage, amortized over four years.

TABLE 4 Cost of Production, Revenue,s and Return on Investment (ROI) for a 48 m^3 Aquaculture Cage in Stratum II of the Volta Lake Stocked at a Density of 125 Fish/m^3 (77.32% Survival Rate) and Cultured for 147 Days (1 US$ = 0.92 Ghana ¢)

	Quantity	Unit Value (GH¢)	Amount (GH¢)
Cost Elements			
Cage (amortized over 4 yrs)	1/2	219.00	109.5
Fingerlings	6000	0.12	720.00
Feed	3000	0.49	1470.00
Labor (pers mos)	6	60.00	360.00
Marketing			50.00
Transportation			100.00
Total cost			**2809.50**
Revenues			
Total harvest (kg)	1176	3.00	3528
Net Income			**718.54**
ROI			**25.6%**

significantly lower in mixed-sex (38.3%) compared to all-male (75.7%) populations. Overall, small fish (<200 g) averaged 17.5 ± 19.73%, medium fish (200–300 g) averaged 27.1 ± 9.9%, and large fish (300–500 g) averaged 54.4 ± 24.1% of the harvest by weight (Table 6). The average price received from fishmongers on the shore within an hour after harvest was ¢3.14 per kg (approximately US $3.00) live weight.

Although water level varied by up to 1.2 m and flow rate was at times nearly undetectable, water quality in the vicinity of the cages (Table 5) was generally stable and remained within the limits for good tilapia growth throughout the trials (Boyd 1990). No fish deaths attributable to poor water quality were recorded. In addition, there was no obvious impact of aquaculture on water quality in the immediate vicinity of the cages (Figure 5).

DISCUSSION

Improper handling and transport of tilapia fingerlings to be stocked in the cages was the major cause of low yields and profits. Typical survival rate in small-scale tilapia cage culture is in the range of 70–80% (Mikolosek et al. 1997, De La Cruz-Del Mundo 1997), although survival as low as 60% has been associated with stocking densities in excess of 70 fish per m^2 (Yi, Kwei Lin, & Diana 1996). In a similar artisanal cage system tested in Côte d'Ivoire by Gorissen (1992), stocking mortality in 30 m^3 cages stocked with 30 g fingerlings at 100 per m^3 was only 5.2%, implying that the problems encountered at Dzemeni can be remedied with proper fish handling techniques, even under rustic conditions. Simple linear regression of the number of marketable fish at harvest against net profit ($y = 1.2x + 2521$; $R^2 = 0.54$) calculated that a farmer using a system similar to that tested at Dzemeni

TABLE 5 Mean ± Standard Deviation of Water Quality Parameters at a Fish Cage Site in Dzemeni on the Volta Lake Measured Bimonthly (n = 10) at Mid-Cage Depth (1 m) for the Period June 2006–December 2007

Temp. °C	pH	Turbidity (NTU)	Transparency (cm)	Electrical Conductivity ($\mu S\ cm^{-1}$)	NO_2-N (mgL^{-1})	NO_3-N (mgL^{-1})	NH_4-N (mgL^{-1})	PO_4-P (mgL^{-1})	DO (mgL^{-1})
30.5 ± 1.30	7.72 ± 0.61	3.28 ± 0.57	140 ± 18.8	59.29 ± 1.627	0.008 ± 0.007	0.41 ± 0.42	0.18 ± 0.14	0.17 ± 0.11	7.90 ± 1.27

TABLE 6 Proportion of Large (>300 g), Medium (250–300 g), and Small (<250 g) Fish Obtained through Six Months of Growout in 48 m³ Cages in the Volta Lake, Ghana

Size category	Percentage Composition Per Cage Per Location						
	Cage 1	Cage 2	Cage 3	Cage 4	Cage 5	Cage 6	Cage 7
Small	49	11	42	7	6	2	13
Medium	32	32	32	42	19	17	16
Large	19	57	26	51	75	81	71

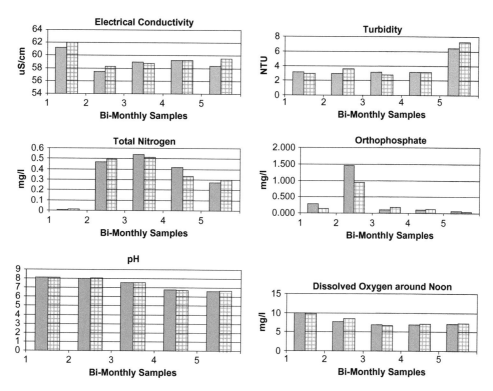

FIGURE 5 Trends in key water quality parameters sampled every eight weeks 10 m above (solid bar on the left) and 30 m below (hashed bar on right) a 192-m³ cage aquaculture facility holding 1.83 MT of fish at final harvest.

would need to produce over 50 fish with an average weight of over 300 g per cubic meter of cage volume to break even.

Typical FCR in *O. niloticus* cage aquaculture systems in Africa is between 1.4 and 2.5 (Beveridge 2004; personal communication, Patrick Blow, Lake Harvest Aquaculture, Zimbabwe, October 2006; personal communication, Steve Murad, Tropo Farms, Ghana, November 2008; personal communication, Karen Veverica, FISH Project, Uganda, January 2009). The higher than usual FCRs realized at Dzemeni were the result of a high percentage of fines in the feed and possible variability in its proximate analysis, coupled to the ±40%

overestimation of the number of fish in each cage as a result of undetected mortality, and thus over-feeding. If the Akosombo strain of *O. niloticus* used in this study has the physiological capacity to achieve a mid-range FCR of 1.6 (Beveridge 2004), then an average of 47% of the feed inputs to the cages was wasted. At an average of 52% of total production costs, a 47% savings in feed would add an additional ¢700 to the bottom line, substantially improving the economics of the system.

In contrast to the findings of Green & Teichert-Coddington (1994), VeraCruz & Mair (1994), Mair et al. (1995), and Kamaruzzaman et al. (2009), but consistent with those of Macintosh et al. (1988), Little, Bhujel, and Pham (2003), and personal communications with commercial cage culturists in Africa (Mark Amechi, Tropo Farms, Ghana, November 2008; Patrick Blow, Lake Harvest Aquaculture, Zimbabwe, April 2009), all-male populations performed significantly ($P < 0.05$) better than mixed sex even up to the relatively modest weight of 300 g. Taken together for purposes of business planning, the overall average specific growth rate of 1.8% body weight per day (mixed sex; virtually unlimited feed) compares favorably with the 1.5% body weight per day calculated from a range of intensive caged tilapia growout trials reported by Balarin & Haller (1982) and El-Sayed (2006), with 1.1 percent realized in low-volume/high-density cages stocked with mixed sex *O. niloticus* in Uganda (Personal Communication, Karen Veverica, FISH Project, Uganda, January 2009) and even the 2.2 percent per day for all male fish produced at Lake Harvest (personal communication, Patrick Blow, Lake Harvest Aquaculture, Zimbabwe, October 2006).

Despite the difficulties encountered, the technical feasibility of the cage culture system was successfully demonstrated. The cages proved sufficiently robust to survive most of the prevailing natural conditions in the Volta Lake. Acts of vandalism and theft are a further risk to which fish cages are sometimes subjected. Prevalence varies with the socio-cultural and economic context, governance, and the type of development, e.g., scale, equitability of benefits sharing, impacts on livelihoods of other stakeholders. If the causes can be understood, it may be possible to mitigate their impact (Beveridge 2004). According to the data collected at Dzemeni, a minimally profitable 48 m^3 small-scale cage aquaculture system in Ghana would have to produce at least 1 ton of fish at an FCR of less than 2.5. With WRI technical assistance, some 20 small-scale investors in the lower Volta River basin are at the time of writing (January 2009) applying the cage aquaculture technology tested at Dzemeni.

During the first trial, 4990 kg of feed were added to the ecosystem, with no detectable effects of cage aquaculture on water quality in the vicinity of the production site. The lack of any clear correlation between water quality parameters that might be expected to fluctuate together (e.g., dissolved oxygen and nitrogen, electrical conductivity, and turbidity) implies that external influences such as currents, localized flooding events, seasonal water level declines, and *inter alia* seem to have an overriding influence on the parameters measured at the production intensity tested. Such observations have often been made in the vicinity of small cage developments. At higher density, cages will undoubtedly have impacts on water quality, indicating the need for careful site selection and ultimately some type of zoning system for cage aquaculture in the Volta Lake and monitoring to support an adaptive management system.

REFERENCES

Abban, E.K., & C.A. Biney. 1996. *Limnological, benthic, macro-invertebrate fauna, and fish biological studies in stratum VII of the Volta Lake in Ghana, Technical Report 157*. Accra, Ghana: Institute of Aquatic Biology.

Abban, E.K., J. Moehl, L.K. Awity, M. Kalende, J.K. Ofori, & A. Tetebo. 2006. *Aquaculture strategic framework*. Accra, Ghana: Ministry of Fisheries.

APHA-AWWA-WEF. 1998. *Standard methods for the examination of water and wastewater*, 20th ed. Washington, DC: American Public Health Association.

Asante, F.A. 2006. Socio-economics of fisheries dependant communities in the Volta Basin of Ghana. *Intl. J. Ecol. Environ. Sci.* 32(1):127–132.

Balarin, J.D., & R.D. Haller. 1982. The intensive culture of tilapia in tanks, raceways and cages. In *Recent advances in aquaculture*, Eds. J.F. Muir & R.J. Roberts, London, UK: Croom Helm. pp. 267–355.

Beveridge, M.C.M. 2004. *Cage aquaculture*, 3rd ed. Oxford, UK: Blackwell Publishing

Boyd, C. 1990. *Water quality in ponds for aquaculture*. Union Springs, AL: Agriculture Experiment Station, Auburn University.

De La Cruz-Del Mundo, R., P. Del-Mundo, M. Gorospe, & R. Macas. 1997. Production and marketing of cage-reared tilapia (Oreochromis niloticus) in Taal Lake, Agoncillo, Batangas. In *Tilapia aquaculture; Proceedings from the Fourth International Symposium on Tilapia in Aquaculture*, November 9–12, 1997, Florida, *Northeast Regional Agricultural Engineering Service, Ithaca, New York*. Ed. K. Fitzsimmons (ed), 633–641.

El-Sayed, A.F.M. 2006. *Tilapia culture*. Wallingford, Oxfordshire, UK: CABI Publishing.

FAO. 2007. Fishstat Plus: a universal database for fisheries statistical time series, V 2.3. Fisheries Department, Food and Agriculture Organization of the United Nations.

Gorissen, G. 1992. Considérations zootechniques et économiques sur l'élevage du *Tilapia nilotica* en cages à Touro (Katiola), République de Côte d'Ivoire. In *Aquaculture systems research in Africa*, Eds. G.M. Bernacsek & H. Powles,. Ottawa, Canada: International Development Research Center., pp. 156–169.

Green, B.W., & D.R. Teichert-Coddington. 1994. Growth of control and androgen-treated Nile tilapia during treatment, nursery and grow-out phases in tropical fish ponds. *Aquacult. Fish. Manage.* 25:613–621.

Hambrey, J. 2006. A brief review of small-scale aquaculture in Asia, its potential for poverty alleviation, with a consideration of the merits of investment and specialization. In *Technical Expert Workshop on Cage Culture in Africa. Fisheries Proceedings 6*. Eds. M. Halwart & J.F. Moehl,. Rome: FAO. pp. 37–47.

Little, D.C., R.C. Bhujel, & T.A. Pham. 2003. Advanced nursing of mixed-sex and mono-sex tilapia (*Oreochromis niloticus*) fry, and its impact on subsequent growth in fertilized ponds. *Aquacult.* 221:265–276.

Macintosh, D.J., T.B. Singh, D.C. Little, & P. Edwards. 1988. Growth and sexual development of 17-methyl-testosterone-and progesterone-treated Nile tilapia (*Oreochromis niloticus*) reared in earthen ponds. In *The Second International Symposium on Tilapia in Aquaculture, ICLARM Conference Proceeding 15*, Department of Fisheries, Bangkok, Thailand and ICLARM, Manila, Phillipines. Eds. R.S.V. Pullin, T. Bhukaswan, K. Tonguthai & J.L. Maclean. pp. 457–463.

Mair, G.C., J.S. Abucay, J.A. Beardmore, & D.O.F. Skibinski. 1995. Growth performance trials of genetically male tilapia (GMT) derived from YY-males in *Oreochromis niloticus* L.: on station comparisons with mixed sex and sex reversed male populations. *Aquacult.* 137:313–323.

Mikolasek, O., J. Lazard, M. Alhassane, P. Parrel, & I. Ali. 1997. Biotechnical management of small-scale tilapia production units in floating cages in Niger River (Niger). In *Tilapia Aquaculture; Proceedings from the Fourth International Symposium on Tilapia in Aquaculture,* November 9–12, 1997, Florida, *Northeast Regional Agricultural Engineering Service, Ithaca, New York*, Ed. K. Fitzsimmons, 348–356.

Kamaruzzaman, N., N.H. Nguyen, A. Hamzah, & R.W. Ponzoni. 2009. Growth performance of mixed sex, hormonally sex reversed, and progeny of YY male tilapia of the GIFT strain, *Oreochromis niloticus. Aquacult. Res.* 40:720–728.

VeraCruz, E.M., & G.C. Mair. 1994. Conditions for optimum androgen sex reversal in *Oreochromis niloticus* (L.). *Aquacult.* 122:237–248.

Yi, Y., C. Kwei Lin, and J. Diana. 1996. Influence of Nile tilapia (*Oreochromis niloticus*) stocking density in cages on their growth and yield in cages and in ponds containing the cages. *Aquacult.* 146:205–215.

Impacts of Aquaculture Development Projects in Western Cameroon

VICTOR POUOMOGNE, RANDALL E. BRUMMETT,
and M. GATCHOUKO

Humid Forest Ecoregional Center, Yaoundé, Cameroon

To measure the impact of past projects on the sustained adoption and development of aquaculture, and to assess the potential for future growth, a participatory rural appraisal (PRA) based on the Research Tool for Natural Resource Management, Monitoring and Evaluation (RESTORE) of 100 farmers (62 with fishponds, 38 without) was undertaken between January and August 2001 in the Noun Division of Western Province, Cameroon. The average household of 14 persons possessed 5.5 ha of land. Educational level is low (less then 35% above primary, 24% illiterate). Most fish producers were small-scale farmers (79%). Of the 360 fish farmers possessing 445 fish ponds (250 m^2 average surface area), only 23% were active. Production is primarily based on earthen ponds stocked with mixed-sex tilapia (Oreochromis niloticus) grown alone (42%) or in polyculture (54%) with the African catfish (Clarias gariepinus). Most ponds are poorly managed, containing underfed fish despite the availability of large quantities of agricultural by-products that could be used as pond inputs. Average annual yield is 1,263 kg/ha. Despite a number of aquaculture development projects over 30 years, there were no significant differences ($P < 0.05$) in household economics and farming systems between fish farming and non-fish farming families. According to active fish farmers, the major constraints to increasing aquaculture production to make it economically interesting are: lack of technical assistance (46%) and lack of good fingerlings (30%). Recent political and economic changes have altered the outlook for *aquaculture in Cameroon, and a development strategy based on new rural development policies is discussed.*

INTRODUCTION

Fish farming as an economic option for small-scale farmers started in Cameroon in 1948 under the French colonial administration. Its subsequent development can be chronologically summarized as follows:

- 1954: Twenty-two public stations are built to strengthen the action of extension agents. More than 10,000 private earthen ponds are built in the country, mostly in the West.
- 1960: Cameroon achieves independence. Aquaculture extension is neglected in favor of coffee and cocoa. Most ponds are abandoned by 1963. Competition with capture fisheries in newly constructed hydroelectric dams reduces the economic viability of aquaculture.
- 1968–1976: A UNDP/FAO regional project increases the number of extension stations to 32. The central research and training station at Foumban trains more than 150 extension agents. However, a financial crisis among donor agencies impedes the completion of the revival project.
- From 1980–1984: USAID and Peace Corps develop common carp culture in the western highlands (Peace Corps volunteers remained the main manpower for aquaculture extension until 1998).
- 1986 to 1990: IDRC supports a project of integrated fish farming.
- 1990–1995: Belgian and Dutch projects on catfish domestication; French assistance to small-scale integrated aquaculture in central region.
- 1990–present: Country-wide agriculture (including aquaculture) support through a national extension and research project financed by the World Bank and the International Fund for Agricultural Development (IFAD), and smaller, more regionally focused general agriculture technical assistance from the Deutsche Gesellschaft für Technische Zusammenarbeit (GTZ).

This aquaculture project history is typical of many African countries and, as in most of these, the results obtained remain far below expectations (Satia, Satia & Amin 1988; Nji & Daouda 1989; Lazard et al. 1991; Hirigoyen, Manjeli, & Mouncharou 1997). For example, in the locality of Foumban where the main Cameroonian aquaculture research and training institutions are located, it is difficult to find a well-managed pond within 50 km.

However, since the economic crisis in 1986, which saw an almost 50% decline in per capita income and precipitated the devaluation of the

Central African Franc (CFA) in 1994, meat consumption by Cameroonian households has decreased in favor of cheaper smoked and frozen fish. To meet the constantly increasing demand for fish in Cameroon, the government imported 54,000 metric tons (MT) in 1997, 64,000 MT in 1998, and 78,000 MT in 1999 of frozen fish, primarily small mackerels (Scombridae) from Senegal and Mauritania. In value, this represents an average of CFA 20 billion (approximately USD $30 million) per year, i.e., second in rank in food imports only to cereals (Oswald & Pouomogne 2000). Along with improving markets, new policies on social organization and personal liberty implemented in December 1992 have helped to revive interest in aquaculture (Oyono & Temple 2003).

This study was conducted to facilitate the formulation of new aquaculture development activities in Cameroon. The main objectives were to measure the impact of past aquaculture projects on farming systems and household economics, and to identify opportunities and options for future work.

MATERIALS AND METHODS

Background knowledge on geography, ecology, agriculture, and demography of the eight subdivisions of the Noun Division of Western Province, Cameroon (Figure 1) was gathered from governmental and non-governmental sources (MINEPIA 2001). A six-person team comprised of two agronomists, three aquaculture technicians, and a senior aquaculture research and extension officer serving as team leader conducted the investigation. Prior to beginning fieldwork, the team received training in participatory methods (including guidance on humility and listening behaviors), and on the use of participatory rural appraisal (PRA) and rapid appraisal of agricultural knowledge system (RAAKS) as surveying tools (Pouomogne 1998).

The investigators, with official documents of introduction from the local research institute, first met the chief of each village or the group leader of the area who directed them to known fish farmers, active or not, where they could collect general biophysical information on the ponds and conduct a brief dialogue with the owner. A total of 360 farmers possessing 445 ponds were thus identified. Data collected included farmer's name and precise location, general information on ponds (e.g., number, size, depth, production system, state of activity), and the farmer's perception of key constraints to aquaculture development in the area.

Following this census, a deeper investigation of the whole farming system was conducted using a comprehensive questionnaire based on the Research Tool for Natural Resource Management, Monitoring and Evaluation (RESTORE), a farming systems analysis software package developed by WorldFish Center (Lightfoot et al. 1993, 1999).

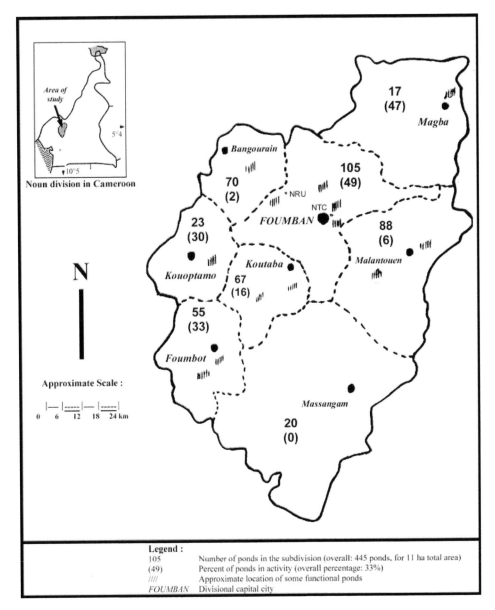

FIGURE 1 Map of Noun Division, Western Province, Cameroon, including number, location, and status of fishponds.

For data collection and analysis, a representative stratified sample of 62 fish-farming households was chosen from the census list. To estimate the impact of fish farming on household economics, this number was supplemented with 38 farmers (also stratified, from all subdivisions) who had never practiced fish farming. Characterization included farm household structure, land/water ownership and management, farming enterprises and other income-generating activities, total farm economy and productivity, and the generation and use of agricultural by-products. This data was supplemented with information related specifically to aquaculture: pond design and sizes, common breeding techniques, harvesting and marketing, technical assistance, and farmers' expectations from research and extension. The application of this questionnaire took an average of 2.5 hours per farmer.

After all 100 farmers had been interviewed, group meetings were organized at four representative subdivisions to synthesize and validate the information gathered. A final meeting took place in the division capital, Foumban, with representatives from each of the eight subdivisions surveyed. Following validation, selected farming systems and household parameters were separately compared using Student's t-test (Zar 1974). The reported value for each of these parameters was calculated as the arithmetic mean of that parameter for the household. Frequencies for a given characteristic were calculated as a percentage of farmers showing that characteristic.

RESULTS

The Noun Division of Western Province, Cameroon (5°45' N – 10°55' E) covers 7,687 km^2 (Figure 1). Most of the land is hilly savanna with wooded valleys. The altitude varies from 500 to 2,270 m (average 1150 m). The climate is unimodal with one long rainy season from April to November, and a dry season from December to March. Annual rainfall averages 1,500 mm, and mean annual temperature is 22°C (minimum 19°C in September, maximum 29°C in March). Evaporative losses from stagnant water bodies during the dry season can reach over 0.8 m. Soils are generally ferrallitic (oxysoils in the U.S. classification), with hydromorphic soils (gleysoils) in swampy areas, and red basaltic soils (cambisoils) in the valley of the Noun River (Pouomogne 1994). Topsoils in the Noun valley are very fertile, with organic matter >4%, pH > 6, a cation exchange capacity (CEC) of >12 meq/100 g, and retainable phosphorous above 10 ppm (Bindzi Tsala & Njomgang 1989). Although temperatures are suboptimal for many tropical fish species, suitable sites for aquaculture are common.

The majority of the Noun's 470,000 inhabitants are small-scale farmers of the Bamoun ethnic group. The second largest ethnic group is the Bororo (a subgroup of the Fulani), who are primarily cattle-herding pastoralists. The rural population density of about 35 inhabitants per km^2 is relatively low.

Household Characterization

There were no significant differences (P < 0.05) between fish farming and non-fish farming households for the parameters measured (Table 1). Almost

80% of the farmers are 30 to 50 years old. Average household size is 14 persons. Just over half of the households are polygamous, with an average of four adult females and seven children. Almost 80% of household members spend more than 60% of their time on agriculture.

Nearly 79% of households farm as their main source of income. Overall, 73% of household revenues are derived from sale of farm produce. Fish accounted for only 0.67% of total income derived from farming. Gross household revenue averages USD $3,857 per year (USD $276 per person). Cash transfers from relatives in urban areas appeared insignificant. Farm production expenses represent 51% of total household expenditures, while purchase of additional food for the family took 18% and health care another 13%.

The annual per capita consumption of animal protein is 15.6 kg of fish (smoked, frozen or fresh), 9.2 kg of beef, 2.2 kg of bushmeat, and 1.9 kg of poultry. Cash expenditures for animal protein average USD $286 per year, equal to 7.4% of total household expenditures.

The Noun Division has the lowest schooling rate in Western Province (UNDP 1999), with 68% of the farmers having no education beyond primary school. According to the farmers themselves, less than 10% of household revenues (i.e., < USD $400 for 7 children per year) are invested in children's education.

All farmers own their land. Farms are usually large (average 5.5 ha), gently sloping, with fertile to highly fertile soils in 90% of cases. Potential areas for fish farming (e.g., wetlands) account for 17% of total land holdings.

Farm Productivity

The main crops produced in Noun Division, in order by weight, are tomatoes, plantains, maize, cassava, and sweet potatoes (Table 2). Average annual production (from mixed crop fields) of the main crops is about half of that reported as typical of African smallholdings by Raemaekers (2001). Maize is the principal staple food crop with an annual per capita consumption of 81.3 kg.

Despite the relatively low yields, an abundance of by-products (Table 3) in the form of weeds, kitchen refuse, and crop residues are produced. Nearly three tons per household of such materials are used primarily as organic fertilizers to mixed-crop food fields or discarded.

Only 3% of fish farmers met their total family fish needs with home production. Other fish farmers and non-fish farmers bought frozen or smoked fish in almost the same quantities. Fresh fish are sold (often on credit that is never paid in cash) on the pond bank (25.4%), eaten by the family (60.4%) or given away as gifts (14.2%). In spite of low yields, more than 56% of fish farmers say they are content with aquaculture as a secondary farming activity.

TABLE 1 Comparison of Socioeconomic Characteristics of Fish Farming vs. Non-fish Farming Households (N = 100 Farms) in Noun Division, Western Province, Cameroon

Parameters	Characteristics	Fish farmers (62)	Non-fish farmers (38)
Age	Below 30	5.00	13.00
	Between 30 and 60	78.50	79.00
	Above 60	16.50	8.00
Religion	Islam	76.70	73.00
	Christianity	23.30	27.00
Marital status	Polygamy	52.50	73.00
	Monogamy	41.40	27.00
	Single (unmarried, widowed, divorced)	5.72	0.00
Membership of a tribal mutual aid association	Yes	47.00	47.00
	No	53.00	53.00
Average number of people in each household	14, of which are:		
	female adults	30.00	30.00
	male adults	25.00	25.00
	female children	23.00	23.00
	male children	22.00	22.00
Percentage of time spent by adults on agriculture	<40%	10.50	22.0
	40 to 60%	10.00	23.00
	>60%	79.50	55.00
Educational level	No formal education	23.60	40.00
	Primary school	44.30	30.00
	Secondary school	30.00	27.00
	Higher education	2.10	3.00
Main occupation	Small-scale Farmer	78.40	90.00
	Other (traditional doctor, trader, baker, teacher, shepherd, smith)	21.20	10.00
Total area of land per household	<4 ha	39.50	57.00
	≥4 ha	60.50	43.00
Land holding right	Exclusive ownership	75.20	80.00
	Mixed ownership and renting	24.80	20.00
Land registration certificate	Yes	18.80	10.0
	No	81.20	90.00
Total area of each farm under various natural resource types*	Homestead	33.40	35.00
	Valleys	11.10	17.00
	Home gardens	10.50	25.00
	Wetlands (swamps)	16.40	5.00
	Fallow	15.90	20.00
Sources of income	Selling farm produce	72.70	73.00
	Remittances, wages or salary	27.30	23.00
Sources of expenditure	Farm operation (seed, tools, fertilizers, pesticides, etc.)	51.00	51.00
	Feeding the family	18.30	18.00
	Education of children	9.65	10.00
	Hired labor	6.20	6.00
	Other (rent, transport, utilities, funerals, medical, etc.)	13.70	13.00

*Terminology after LIGHTFOOT et al. (1993, 1999).

TABLE 2 Production and Consumption of Most Important Food Items by Rural Households of Noun Division, Western Province, Cameroon. Average farm size = 5.5 ha

	Production		Purchased		Consumption	
	Rank	Quantity (kg/farm/year)	Rank	Quantity (kg/farm/year)	Rank	Quantity (kg/farm/year)
FOOD CROPS						
Plantain *Musa sapientum*	1	1895	2	227	2	927
Maize *Zea mays*	2	1773	1	231	1	1138
Cassava *Manihot esculentum*	3	1142			3	657
Sweet potato *Ipomoea batatas*	4	1006			4	384
Groundnuts *Arachis hypogea*	5	384	5	33	5	240
Rice *Oriza sativa*			3	91		
Beans *Phaseolus* spp.			4	77		
FRUITS AND VEGETABLES						
Avocado *Persea americana*			5	26		
Palm oil *Elaeis guineensis*			1	90	3	185
Condiments (leek, garlic, parsley)			3	67	4	170
Banana *Musa paradisica*			2	80	1	299
Tomato *Lycopersicon esculentum*	1	2979				
Peppers *Capsicum* spp.	2	722				
Mango *Mangifera indica*	3	693				
African plum *Dacryodes edulis*	4	678				
Black Nightshade *Solanum nigrum*	5	547	4	65	2	198
ANIMAL PRODUCTS						
Fish (72% smoked, 28% frozen)			1	191	1	193
Beef	1	12	2	117	2	129
Fresh fish	2	10	4	23	5	25
Poultry (chicken and ducks)	3	9	5	20	4	26
Goats and sheep	4	6				
Bushmeat	5	1	3	30	3	31

No significant differences ($P < 0.05$) were found between fish farmers and non-fish farmers for any non-aquaculture aspects of the farming system.

Aquaculture

Among the 62 fish farmers, there were 77 ponds (Table 4). Of these, 18 are still functioning, the remaining 77% being abandoned. The majority (60%) of

TABLE 3 Agricultural and other By-products Produced on Small-scale Farms in Noun Division, Western Province, Cameroon. Dry matter (DM) and Nitrogen Estimates Based on Göhl (1992)

By-product	Part used	Current use	DM (%)	Nitrogen (% DM)	Quantity (kg/farm/year)
Weedy grasses*	Whole	Crop Fertilization	25	≤ 3.2	> 5000
Rotten fruits & various leaves		Crop Fertilization	20	≤ 3.2	> 500
Harvest residue		Crop fertilization, animal feed	50	0.6	> 2000
Animal droppings		Crop and pond fertilization	60	≤ 3.2	> 650
Kitchen waste		Refuse or fishpond	22	3.0	> 1300
Wood ash		None			> 400
Maize (*Zea mays*)	Spoiled grain	Animal feed	89	1.4	80
	Bran	Animal feed	93		234
	Stovers	None or crop fertilization	89		> 1500
Cassava (*Manihot esculentum*)	Leaves	Human food	16	0.6	> 500
	Peels	None			
Sweet potato (*Ipomoea batatas*)	Leaves	None	11	1.0	> 200
Papaya (*Carica papaya*)	Leaves	None	22	3.9	> 100
Plantain (*Musa* spp.)	Leaves & peels	None	94	1.3	> 200
Rice (*Oriza sativa*)	Bran	Crop & pond fertilization	89	3.1	60
Oil Palm (*Elaeis guineensis*)	Kernels	Human food, Fuel	85	4.3	
	Nut pulp	None or fuel			
Coffee (*Coffea arabica, C. robusta*)	Pulp	None or crop fertilization	94	1.6	

Chromolaena odorata, Pennisetum purpureum, Imperata cylindrica, Ageratum conyzoides.

ponds (both active and inactive) were constructed prior to the 1960s. Most are poorly designed and minimally managed. Ponds tend to be shallow, with little or no bottom slope and narrow dikes. Although over 90% have reasonable water retention, a large number are undrainable (77%), full of weeds (24%), and/or less than 50 cm deep (20%).

Table 5 describes the management of active fishponds in Noun Division. About half of fish farmers produce tilapia (*Oreochromis niloticus*) in monoculture, while the others grow a polyculture of tilapia and catfish (*Clarias gariepinus*). Despite the availability of abundant agricultural by-products that could be used as inputs, ponds tend to be under-fertilized (67%). Compounding the problem of low inputs, the majority of fish farmers have overstocked or lost track of how many fish might be in their ponds.

TABLE 4 History and Characteristics of Fishponds in the Noun Division, Western Province, Cameroon

Parameters	Characteristics	Percent of sample
Type of pond	Derivation (contour) pond	90.00
	Barrage (dam or embankment) pond	10.00
Number of ponds	77 total ponds of which farmers with:	
	1 pond only	42
	2–3 ponds	47
	>3 ponds	11
Dike top width	< 50 cm	23.61
	50–100 cm	59.16
	>100 cm	17.23
Pond depth	<50 cm	24.20
	50–100 cm	52.38
	>100 cm	23.42
Pond bottom design	Without slope	67.17
	Irregular	10.09
	Sloped towards the drain	22.74
Date of pond construction	Before 1960	60.00
	Between 1960–1990	31.11
	After 1990	8.89
Ponds in Use	Active	77.0
	Abandoned	23.0
Motive for pond construction	Family food	50.00
	Income	33.33
	Recreation	16.17
Major reason given for abandonment	Lack of capital	30.00
	Lack of labor	20.00
	Lack of technical support	14.78
	Low profits relative to coffee	11.11
	Lack of fish seed	8.85
	Other (pond flooding/drying, death of owner, competition from new dams)	15.26

Almost all fish farmers (92%) harvest their ponds once a year at festive periods. In 41% of cases, intermittent partial harvesting by hook and line angling also takes place. Most ponds (89%) are drained by breaking the dike.

The quantity of fish produced per household is only 1,263 kg/ha/yr (31.6 kg for the typical family pond of 250 m^2) for the 33% who are active fish farmers, and 417 kg/ha/yr overall (10.4 kg per pond). This is low in comparison to the 3–8 MT/ha/yr estimated as possible by Pouomogne, Nana, and Pouomegne (1998). According to active fish farmers, the main constraints to increasing production are: lack of technical support (35%), lack of fish seed (20%), predation and theft (14%), and quality of fish seed (5%).

TABLE 5 Fishpond Management in the Noun Division, Western Province, Cameroon

Parameters	Characteristics	Percent of sample
Pond general appearance	Clean	75.68
	Little bush	24.32
Water control/harvesting method	Earthen canals/dike breaking	89.40
	PVC standpipes with elbow/complete draining	10.60
Water level in the pond	Very shallow	20.00
	Half full	45.71
	Full	34.29
Pond fertilization state	Fertile (green water)	33.33
	Not fertile (transparent water)	66.67
Presence of a composting system	Yes	19.78
	No	80.72
Species in culture	*O. niloticus* + *C. gariepinus*	53.71
	O. niloticus only	42.16
	Other (*Barbus, Hemichromis*)	4.13
Fish stocking rate	< 2 per m^2	20.00
	> 2 per m^2 or unknown	80.00
Fish seed origin	Public station	81.40
	Wild caught or other fish farms	18.60
Inputs per 250 m^2 pond per week	An average of 27.4 kg of which:	
	Maize bran and spoiled grain	40.15
	Animal droppings	34.31
	Kitchen refuse	18.25
	Other (fruits, leaves, grasses, cassava soaking)	7.29
Number of times food/fertilizer applied per week	1	12.56
	≥ 2	87.44
Type of labor	Family	70.00
	Family + hired	30.00
Intermittent angling (partial harvest)	Yes	41.00
	No	59.00
Production cycle	6–12 months	92.00
	> 12 months	8.00
Mean annual yield (32 kg per 250 m^2)	An average of 1263 kg/ha/yr of which:	
	< 750 kg/ha/yr	40.00
	750–1500 kg/ha/yr	10.00
	>1500 kg/ha/yr	50.00
Disposal of produce	Home consumption	60.42
	Sale	25.38
	Gifts	14.20
Farmers impression	Satisfied	56.00
	Not satisfied	44.00
Price of pond fish (per kg)	*C. gariepinus*	600–1200 CFA
	O. niloticus	500–1000 CFA
Major problems felt by farmers	Lack of technical support	35.00
	Lack of fish seed	20.00
	Predation and thefts	14.00
	Quality of fish seed	5.00
	Others (pond flooding/drying, marketing)	26.00

DISCUSSION

Assuming that there were no *a priori* differences between fish farming and non-fish farming families in terms of the household and farming-system parameters sampled in this study (unless aquaculture actually had a negative impact), it seems clear that there were no lasting improvements in household economic status accruing to farm families adopting aquaculture in the area studied. Nevertheless, a majority of active fish farmers were content with their ponds.

To understand this seeming contradiction, one needs to look at the role a fishpond typically plays in an African farming system. As a secondary activity, after staple crop production (maize and plantains, in the case of the Noun), a fishpond is similar to chickens, goats, vegetable production, and a number of crops grown primarily for cash (e.g., tobacco, cotton, coffee, tea, cacao) depending upon the climate. Small-scale African farming systems produce dozens of such crops, nearly 130 having been identified by Dupriez & De Leener (1988). The majority of these products, including fish, are either consumed by the family or traded to neighbors and/correlatives (Brummett 2000). Surpluses and crops grown exclusively for cash are sold. Rather than making tradeoffs and taking risks by allocating all of the farm resources to the one or two most profitable of these, small-scale farmers tend to diversify by growing a number of crops simultaneously (often in mixed plots), thereby lowering overall risk. This can be an important, even lifesaving, strategy in low external input rain-fed agriculture. However, the intended result of the aquaculture projects implemented in Cameroon was not diversification of farming systems, as logical and beneficial as that may be, but changes in the economic status of fish farming families. This was clearly not achieved.

Although the objectives of this study did not include a systematic review of past aquaculture projects in western Cameroon, several characteristics of these are typical of African aquaculture projects in general and may have contributed to the low level of sustainable impact (Brummett et al. 2004):

- Focusing on fish in isolation from other farming activities and social/economic relationships within the community. In effect, projects did not take into consideration the full range of non-technical constraints faced by smallholding farmers.
- A generally top-down relationship between the donor, research, extension, and farmers that effectively prevented joint learning and concentrated effort on the donor's priorities.
- Poorly trained extension personnel who neither understood the technologies they were promoting nor could communicate effectively with small-scale farmers.

In addition, most of these projects received substantial external subsidization. Fish hatcheries and technical assistance services were supported either directly or through foreign assistance programs. With donor fatigue and structural adjustment in the early 1990s, the government could no longer afford this support. Lack of technical oversight and limited (and high-priced) seed of mixed quality led to low yields and profits. Labor and capital were consequently transferred to more profitable farm crops such as maize, tomatoes and, until recently, coffee.

Since these projects were implemented, economic and political changes have altered the prospects for rural development in Cameroon (Oyono & Temple 2003). The price of fresh pond fish (500 to 1200 CFA/kg) is competitive with other fish products (including fresh from natural water bodies, smoked or frozen). Fresh fish at lower prices can only be found in the vicinity of substantial capture fisheries. In addition, population has increased and the demand for food, especially animal protein, has increased accordingly. Current average annual demand for fish in the Noun is over 15 kg/person, compared to 9 kg for beef and 2 kg each for bushmeat and poultry. With production of only 0.9 kg/person/year from aquaculture and capture fisheries static or declining, fish prices can be expected to increase further.

CHALLENGES FOR RESEARCH AND EXTENSION

Western Cameroon appears to have high biophysical potential for aquaculture. However, several key constraints have prevented fish farming from increasing the economic status of fish farmers. From the findings of our survey, the main challenges to research and extension appear to be:

- Transfer of existing technical knowledge to poorly educated farmers.
- Availability of high-quality fingerlings (seed).

These problems are not really new or surprising. They are probably common to the majority of agricultural innovations that could be profitably adopted by African small-scale farmers. The fact that they are common, however, does not mean that their solution will be simple.

At present, the majority of the most suitable aquaculture technology in Africa is known only within the research community. Unfortunately, researchers are not commonly in direct contact with farmers and interact with extension agents only through short training and discussion seminars. Efforts are needed to bridge the information gap between those who study and develop technology and those who use it. Taking into account the low government budgets for support to aquaculture, the existing (and expensive) Training and Visit system cannot be considered as the best alternative for meeting the needs of farmers. Regardless of the dissemination system selected, changes in the policies of government aquaculture research and extension services will not be effective unless attached to substantial changes in financing.

Shortage of fingerlings is a long-standing problem for African fish farmers. A number of strategies have been adopted to address this difficulty. In many African countries, donor-assisted projects built large government hatcheries that were supposed to produce subsidized fingerlings for small-scale producers. This strategy largely failed, for the same reasons mentioned above, for African aquaculture projects in general.

More recently, a number of agencies have successfully developed hatchery systems for smallholders themselves, and in a number of places over 80% of fingerlings stocked are purchased from small-scale hatcheries. This, however, has created another problem: genetic erosion due to poor broodstock management. In a typical chicken-or-egg conundrum, small-scale hatcheries cannot flourish in the absence of substantial numbers of growers, while growers cannot prosper without adequate numbers of good fingerlings. The result has been a large reduction in genetic quality of fingerlings. Eknath and colleagues (1993) and Brummett, Etaba Angoni and Pouomogne (Forthcoming) found that wild tilapia (*Oreochromis niloticus*) populations grow some 40% faster than populations held by small-scale hatcheries. This observation is supported by anecdotal evidence from Uganda (R.Gregory, Kajjansi Aquaculture Research and Development Centre, personal communication) and Zimbabwe (P. Blow, Lake Harvest Aquaculture Ltd., Kariba, personal communication), among others.

From the authors' experience, what is required to overcome the gap in know-how between research and farmers (both growers and hatchery operators) is an approach that permits joint learning and participatory technology development among farmers and researchers together. A dynamic and direct relationship between research and farmers has been shown to produce substantial positive impacts in the aquaculture sectors of industrialized countries (Brummett 2003), and such an approach might be adaptable to the African context to maximize the impact of limited aquaculture development spending.

Another strategy that could be particularly useful in overcoming the fingerling quantity and quality problems is a new institutional arrangement among private and public sector agencies. Depending upon local economic conditions, larger-scale operators might be able to support sufficiently large investments to permit the proper genetic management of broodfish populations. Rather than being excluded as "too wealthy" by donor agencies, these investors could be brought into the rural development and extension process as breeders and hatchery operators. Cash flow generated by other investments would enable larger-scale operators to suffer the gap between availability of fingerlings and expansion of grow-out farms. Since more aquaculture means more potential fingerling sales, hatchery operators would also have an incentive to provide technological advice to farmers (Lewis, Wood, & Gregory 1996). Albeit without sufficient regard for genetic quality, Madagascar, with the assistance of the Food and Agriculture Organization of the United Nations, used a medium-scale hatchery-led model of this general type to advantage in the development of an aquaculture industry that grew from 160 MT prior to the project to over 6,000 MT per annum valued at more than USD $7 million within six years (Shatz 2000). Such an approach, combined with the inclusion of somewhat larger hatcheries that could maintain and even improve the quality of seed through selective breeding, might overcome one of the most serious and long-standing constraints to African aquaculture.

REFERENCES

Bindzi Tsala, J. and R. Njomgang. 1989. *Caractérisation morphologique, physique et chimique des sols à hydroxydes dans trois toposéquences de l'Ouest-Cameroun.* Mémoires et Travaux de l'Institut de Recherche Agricole 3. Yaoundé. Cameroun: Ministry of Research.

Brummett, R.E. 2000. Factors affecting fish prices in southern Malawi. *Aquaculture* 186(3/4):243–251.

Brummett, R.E. 2002. Seasonality, labor, and integration of aquaculture into Southern African smallhold farming systems. *Naga, The ICLARM Quarterly* 25(1):23–27.

Brummett, R.E. 2003. Aquaculture and society in the 21st century. *World Aquacult.* 34(1):51–59.

Brummett, R.E., V. Pouomogne & A.G. Coche. 2004. Aquaculture extension in sub-Saharan Africa. Extension Circular 1002. Food & Agriculture Organization of the United Nations, Rome.

Brummett, R.E., D. Etaba Angoni, & V. Pouomogne. Forthcoming. On-farm and on-station comparison of wild and domesticated Cameroonian populations of *Oreochromis niloticus. Aquaculture.*

Dupriex, H., & P. De Leener. 1988. *Agriculture in African rural communities.* London: Macmillan Education Ltd.

Eknath, A.E., M.M. Tayamen, M.S. Palada-de Vera, J.C. Danting, R.A. Reyes, E.E. Dionisio, J.B. Capili, H.L. Bolivar, T.A. Abella, A.V. Circa, H.B. Bentsen, B. Gjerde, T. Gjedrem, & R.S.V. Pullin. 1993. Genetic improvement of farmed tilapias: the growth performance of eight strains of *Oreochromis niloticus* tested in different farm environments. *Aquaculture* 111:171–188.

Göhl, B. 1992. *Tropical feeds: Software developed by Oxford Computer Journals.* Rome, Italy: FAO.

Hirigoyen, J. P., Y. Manjeli, & G. C. Mouncharou, 1997. Caractéristiques de la pisciculture dans la zone forestière du Centre Cameroun. *Tropicultura* 15(4):180–185.

Lazard, J. Y. Lecomte, B. Stomal, and J-Y Weigel. 1991. *Pisciculture en Afrique subSaharienne.* Paris: Ministère de la Coopération et du Développement.

Lewis, D., G. Wood, & R. Gregory. 1996. *Trading the silver seed: Local knowledge and market moralities in aquaculture development*. Rugby, UK: Intermediate Technology Development Group Publishing.

Lightfoot, C., J.P.T. DaIsgaard, M.P. Bimbao, & F. Fermin. 1993. Farmer participatory procedures for managing and monitoring sustainable farming systems. *J. Asian Farming Systems Assoc.* 2(2):67–87.

Lightfoot C., M.P. Bimbao, T.S. Lopez, F.D. Villaneuva, E.A. Orencia, J.P.T. Gayanilo Jr., M. Prein, & H.J. McArthur. 1999. *Research tool for natural resource management, monitoring and evaluation (RESTORE): Field guide. ICLARM Software 9*. Penang, Malaysia: WorldFish Center.

MINEPIA. 2001. *Annual report, Noun Division*. Foumban, Cameroon: Ministry of Animal Husbandry and Fisheries.

Nji, A. & Daouda. 1990. Facteurs techniques lies a l'abandon de la pisciculture dans les provinces de l'Ouest et du Nord Ouest. *Tropicultura* 8(4):189–192.

Oswald, M., & V. Pouomogne. 2000. *Etude de faisabilté pour un développement de la pisciculture villageoise dans les provinces du Centre et de l'Ouest du Cameroun. Rapport définitif*. Yaoundé, Cameroun: APDRA-F/MCAC.

Oyono, P-R., & L. Temple. 2003. Métamorphose des organisations rurales au Cameroun: Implications pour la recherche-développement et la gestion des ressources naturelles. *Revue Internationale de L'Economie Sociale* 288:68–79.

Pouomogne V. 1993. Growth response of *Oreochromis niloticus* to cow manure and supplemental feed in earthen ponds. *Rev. Hydrobiol. Tropics* 26(2):153–160.

Pouomogne V. 1994. Alimentation du tilapia *Oreochromis niloticus* en étang: Evaluation du potentiel d'utilisation de quelques sous-produits de l'industrie agro-alimentaire et modalités d'apport des aliments. Thèse de Doctorat d'Halieutique, ENSA de Rennes. Edité par le CIRAD, Montpellier, France.

Pouomogne V. 1998. *La (RAD) Recherche Agricole orientée vers le Développement, Version ICRA: Que peut en tirer l'IRAD du Cameroun?* Montpellier, France: IRAD Yaoundé/ICRA

Pouomogne, V., J-P. Nana, & J.B. Pouomegne, 1998. *Principes de pisciculture appliqués en milieu tropical africain. Comment produire du poisson à coût modéré. (Des exemples du Cameroun)*. Yaoundé, Cameroun: CEPID-Cooperation Française, Presses Universitaires d'Afrique.

Raemaekers, R.H. (Ed.) 2001. *Crop production in tropical Africa*. Brussels, Belgium: Directorate General for International Cooperation, Ministry of Foreign Affairs, External Trade and International Cooperation.

Satia, B.P., P.N. Satia and A. Amin. 1991. Large scale reconnaissance survey of socioeconomic conditions of fish farmers and aquaculture practices in the West and Northwest provinces of Cameroon. In *Aquaculture systems research in Africa, IDRC-MR 308e, f*, G.M. Bernaseck and H. Powles (Eds.), 64–90. Ottawa, Canada: International Development Research Centre.

Shatz, Y. 2000. *FISHSTAT electronic database*. Rome, Italy: FAO.

Zar, J.H. 1974. *Biostatistical analysis*. Englewood Cliffs, NJ: Prentice-Hall.

Rearing Rabbits Over Earthen Fish Ponds in Rwanda: Effects on Water and Sediment Quality, Growth, and Production of Nile Tilapia *Oreochromis niloticus*

SIMON RUKERA TABARO[1,3], ONISIMO MUTANGA[1], DENIS RUGEGE[1], and JEAN-CLAUDE MICHA[2]

[1]*University of KwaZulu-Natal, Faculty of Sciences and Agriculture, School of Environmental Sciences, Geography, Centre for Environment, Agriculture & Development, Pietermaritzburg, South Africa*
[2]*University of Namur, Department of Biology, Research Unit in Environmental Biology, NAMUR, Belgium*
[3]*National University of Rwanda, Faculty of Agriculture, Department of Animal Productions, Butare, Rwanda*

Nine earthen ponds of 400 m^2 each, were stocked with 800 mixed sex Nile tilapia Oreochromis niloticus *fingerlings (14 g mean weight stocked at 2 fish/m^{-2}) and fertilized with rabbit droppings from rabbits reared over fish ponds and stocked at three different densities: T1 = one, T2 = two, and T4 = four rabbits per 100 m^2 of pond. After 152 days, results from this integrated rabbit/fish system showed: 1) increasing nutrient content of all ponds in the three treatments except for nitrates, which decreased with time and, 2) good water quality in terms of pH, dissolved oxygen, and turbidity. Fish mean weight at harvest and fish yield were higher in ponds fertilized by the highest rabbit stocking rate: 42.32 g and 6.35 ± 1.0 kg/are, respectively (1 are = 100 m^2).*

This work is the first part of an ongoing PhD research funded by the Nile Basin Initiative (NBI-ATP) to whom I would like to convey my gratitude for assistance.

INTRODUCTION

Fish production in Rwanda is dominated by Nile tilapia (*Oreochromis niloticus*). Fish farming is undertaken in individual and cooperative fishponds that are sometimes integrated with crop production and to a lesser extent with livestock production. Ponds are fed occasionally and fertilized using different types of grass compost cut from the pond dikes and very small amounts of animal manure produced on hills near homesteads where animals are kept (Micha 2001).

Rabbits are herbivorous animals consuming high roughage and offering limited competition to humans and other domestic animals for similar foods. Rabbits utilize fibrous by-products that are not useful for poultry or swine. Rabbit manure is a rich source of nitrogen and phosphorus (Franco 1991; Kumar & Ayyapan 1998). Lebas et al. (1996) and McCrosckey (2001) reported NPK content of rabbit droppings in the ranges of 2.4–3 N to 1.4–1.5 P and 0.3–0.6 K richer than other farm manures.

During his study on the effect of rabbit dung on the growth of *Oreochromis niloticus*, Breine et al. (1996) observed a fish growth rate 5.1 times that of the control group. Results from Cameroon noted a positive effect of rabbit manure as a water fertilizer for *O. niloticus* ponds (Breine, Teugels, & Ollevier 1995). Rearing rabbits over fish ponds was tried for the first time in Rwanda in 1988 by researchers of the National Fish Culture Service (SPN) at Kigembe (Southern Province), where Van Vleet (1997) reported similar fish yields (6,000 kg/ha) to those obtained from integrated chickens (6,500 kg/ha), pigs (5,000 kg/ha), and ducks (4,500 kg/ha), and recommended a loading rate of 60 rabbits per 400 m^2 of pond surface, but did not make any observation on the impact of rabbit dung on water quality.

This study is a start of part of a broad research program to investigate the potential of the integration of agriculture and aquaculture in Rwanda. It focuses on the effects of rabbit droppings on fish pond water and pond sediment quality and consequently on the growth and production of Nile tilapia.

MATERIAL AND METHODS

The experiment was conducted between March 20 and August 19, 2008, in nine earthen ponds (400 m^2, 110 cm maximum water depth) at the Rwasave Fish Farming Research Station of the National University of Rwanda, Butare, Rwanda (geographic coordinates: 2°40′S and 29°45′E). Rabbits were reared in hutches built over fish ponds that were first drained, dredged, dried, and then refilled with water from an open supply canal. After the initial filling, water was added only to compensate for losses from evaporation and seepage. Three treatments were assigned randomly in triplicate ponds fertilized with rabbit droppings falling directly from 1, 2, and 4 rabbits per

100 m² of pond, being treatments T1, T2, and T4. Rabbits were fed grass cut from the pond levees *ad libitum* and received water *ad libitum*. Each pond was stocked with 800 fingerlings of mixed sex *O. niloticus* (average weight: 14.9 g) at a density of 2 fish per m². No supplementary feed was given to the fish during this experiment.

Water transparency (Secchi disk visibility), temperature, dissolved oxygen (WTW OXI 325), pH (WTW pH320), and electric conductivity (WTW LF 38) were measured every two weeks at 07:00 a.m. directly in ponds at the surface (~10 cm depth) and bottom (~85 cm depth). At the same time, water samples were collected at depths of between 70 and 100 cm in the middle of each pond using a Van Doorn water sampler were mixed and kept at 4°C for laboratory analysis of nutrient content. Ammonia nitrogen was analyzed following the method of Descy (1989), nitrate nitrogen was analyzed according to Boyd (1979), and nitrites were analyzed using standard methods (Franson et al. 1985). Total nitrogen was analyzed according to Léonard and Kanangire (1998). Orthophosphate was analyzed by the method used by Descy (1989), while total phosphorus was determined according to Léonard and Kanangire (1998). Total alkalinity was determined using the method of Descy (1989). Primary productivity was determined according to Dauta and Capblancq (1999).

Pond sediment nutrients were assessed in eight samples per pond. Total organic nitrogen was determined by standard methods (Franson et al. 1985). Total phosphorus was measured according to IITA (1975) adapted by Léonard and Kanangire (1998). Orthophosphate was determined with the method of Pages et al. (1982) as cited by Léonard and Kanangire (1998).

All the rabbits and 35 fish per pond were sampled every two weeks, weighed on an electronic balance, and returned to the pond. Ponds were completely drained 127 days after stocking and the fish harvested, counted, and weighed to the nearest gram. Fish data at stocking and harvest were used to calculate survival rate: SR (%) = $100 - (100/[N1-N2]/N1)$, the specific growth rate SGR (%/d^{-1}) = $100*([\ln W2 - \ln W1]/d2 - d1)$, and daily weight gain DWG (g/d^{-1}) = $(W2-W1)/(d2 - d1)$, where N = Δt (the rearing time in days), net production (kg/a/yr), and yield (kg/a). Significant differences among treatments was determined via one-way ANOVA and the least significant difference (LSD) test.

RESULTS

Rabbit average body weight increased in all treatments, moving from 500 to about 2,000 g. Rabbits at one per square meter (T1) gained significantly ($P < 0.05$) more weight than the other densities (Figure 1). Rabbits in T1 gained 11.4 g/day^{-1}, while rabbits T2 and T4 gained an average of 9.0 g/day^{-1}.

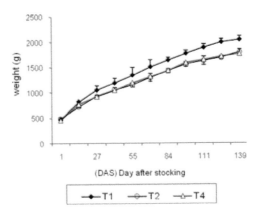

FIGURE 1 Rabbit weight gain when reared over fish ponds for 152 days at 1, 2, and 4 rabbits per square metre hutch.

Mean fish weight was not significantly different among treatments ($P < 0.05$) at the end of the experiment (Table 1). DWG (0.10 to 0.15 g/d) and fish yield (4.61 and 6.35 kg/are) obtained in this study were low, corresponding to an extrapolated annual production of between 11 and 18 kg/are/year. The treatment that received droppings from the higher rabbit stocking rate (T4) produced higher SGR and fish yield ($P < 0.05$) than T1 and T2. Survival did not differ among treatments.

Overall, water quality remained within safe limits for tilapia production (Table 2). Mean minimum concentrations of DO measured in the morning ranged between 1.41 and 1.67 mg/l. A significant decrease in morning dissolved oxygen was observed in all treatments (Figure 2). A similar trend was evident in water transparency (secchi disk visibility, Figure 3). Primary productivity increased to day 57 and then decreased (Figure 4).

Total alkalinity (mg/l $CaCO_3$) in the pond water did not differ ($P = 0.06$) among treatments (Figure 5). Slightly higher values were observed in ponds receiving droppings from a larger number of rabbits. Mean alkalinity varied between 43.33 and 72.67 mg.l^{-1} with a trend of increasing alkalinity over time seen mostly in T4. At the beginning of the experiment, water conductivity was between 105 and 122 µS/cm and remained about the same except for in treatment T4, which had the most rabbits (Figure 6). Electrical conductivity in T4 (158 ± 2.5 µS/cm) was significantly higher than T1 (128.7 ± 1.1 µS/cm) and T2 (132.5 ± 0.6 µS/cm).

Levels and trends in pond water nutrient (N and P) concentrations are shown in Table 3 and Figure 7. Ammonia increased significantly ($P < 0.05$) with time in all treatments. The treatment with more rabbits (T4) had a significantly higher ($P < 0.001$) ammonia concentration than others. Mean ammonia concentrations overall were between 0.33 ± 0.2 mg/l and 1.07 ± 0.5 mg/l. Nitrite concentrations were significantly higher ($P < 0.001$) in ponds treated with four rabbits per are. Mean nitrite concentrations were

TABLE 1 Growth and production performance of Nile tilapia (*O. niloticus*) reared in earthen ponds fertilized by rabbit dung at three different densities for nearly five months. Values are average ± std error of the mean. Data with the same superscript in the same row are not significantly different (p < 0.05)

Treatment	Number stocked	Mean Wt (g) Day$_0$	Surv. (%)	Mean wt (g) Day$_{127}$	Yield (kg.are^{-1})	DWG (g.day^{-1})	SGR (%)	SGR (%/day^{-1})
T1	707	13.93 ± 0.49	80	31.65 ± 6.5	5.14 ± 0.78^a	0.10 ± 0.02^a	79.82 ± 4.33^a	0.43 ± 0.15^a
T2	753	13.24 ± 0.61	89	29.49 ± 0.59	4.61 ± 0.44^a	0.14 ± 0.01^a	85.19 ± 6.54^a	0.58 ± 0.04^b
T4	800	14.74 ± 0.40	93	42.30 ± 2.73	6.35 ± 0.44^a	0.15 ± 0.00^a	93.38 ± 3.06^a	0.77 ± 0.04^c

TABLE 2 Means ± standard errors (and data range) in water quality at three different rabbit stocking rates (1, 2, 4 rabbits/are) in a rabbit/fish integrated production system. Values represent an average of surface and bottom layers measurement over the entire 152 days production period. There were no significant different in any parameters among treatments ($P < 0.05$)

Parameter	Treatments		
	T1	T2	T4
Temperature (°C)	22.03 ± 0.20	21.97 ± 0.20	21.96 ± 0.19
	(20.09–24.42)	(20.06–24.86)	(20.2–25.26)
pH	6.83 ± 0.19	6.53 ± 0.22	6.69 ± 0.26
	(5.57–8.81)	(5.0–8.87)	(5.55–9.08)
Conductivity ($\mu S.cm^{-1}$)	115.64 ± 1.48	120.45 ± 0.65	137.89 ± 3.37
	(104.83–128.0)	(110.75–132.50)	(111.80–155.67)
Dissolved Oxygen ($mg.L^{-1}$)	5.0 ± 0.38	4.71 ± 0.43	4.92 ± 0.60
	(1.67–8.91)	(1.59–8.17)	(1.41–10.48)
Primary productivity ($mgC.m^{-2}.day^{-1}$)	296.39 ± 25.8	321.28 ± 29.2	292.46 ± 22.6
	(178.01–353.22)	(238.59–426.82)	(243.55–389.05)
Secchi disk visibility (m)	0.39 ± 0.01	0.36 ± 0.01	0.31 ± 0.01
	(0.33–0.45)	(0.32–0.41)	(0.23–0.43)

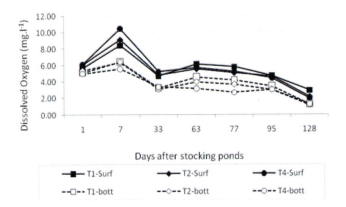

FIGURE 2 Trend in oxygen concentration in tilapia ponds with three different rabbit densities (1, 2, and 4 rabbits/are) to provide nutrient inputs. Each value is an average of three measurements at different points in the pond taken at the surface (solid line) and bottom (dashed line).

0.70 ± 0.10 mg/l in T4, while concentrations were 0.40 ± 0.09 mg/l and 0.50 ± 0.12 mg/l in T1 and T2, respectively. Total nitrogen (TN) concentration increased over time ($P < 0.001$) starting at an average of 0.04–0.06 mg/l and ending between 0.10–0.14 mg/l at the end of the experiment, with no significant differences among treatments ($P < 0.05$). Nitrate concentrations were 0.75 ± 0.51, 1.46 ± 0.43, and 2.72 ± 0.60 mg/l, respectively for T1, T2, and T4. Treatment T4 had significantly higher concentrations of nitrates-nitrogen ($P < 0.001$), while no statistical difference was evident between

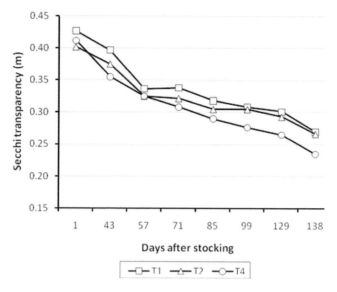

FIGURE 3 Water transparency (secchi disc visibility) in tilapia ponds over 152 days of rearing rabbits over fish ponds at three different rabbit densities (1, 2, and 4 rabbits/are).

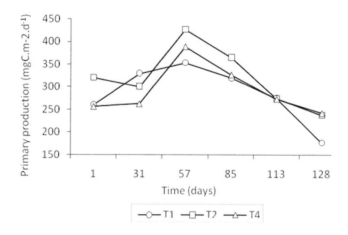

FIGURE 4 Primary productivity in fish ponds fertilized by rabbit droppings for 152 days.

T1 and T2. Phosphorus showed a similar trend to nitrogen, increasing significantly over time ($P < 0.001$) in all treatments with the most number of rabbits (T4) producing significantly higher phosphorus concentrations. Mean total phosphorus concentrations over the entire production period were 0.62 ± 0.11, 0.69 ± 0.12, and 0.94 ± 0.10 mg/l, respectively, in T1, T2, and T4, and orthophosphate 0.07 ± 0.02, 0.10 ± 0.02, and 0.21 ± 0.07 mg/l, respectively in T1, T2, and T4.

Bottom soil of ponds receiving rabbit droppings were significantly enriched in total phosphorus and total nitrogen (Figure 8). The magnitude

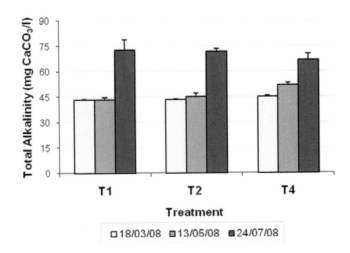

FIGURE 5 Average total alkalinity (mg/l CaCO$_3$) of fish pond water fertilized with three different rabbit densities over 152 days.

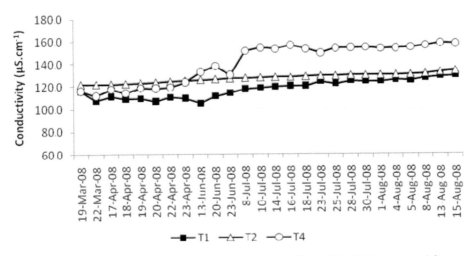

FIGURE 6 Trend in water electrical conductivity (μS.cm^{-1}) in rabbit/fish integrated farming (T1 = 1 rabbit/are; T2 = 2 rabbits/are, and T4 = 4 rabbits/are). Each value is an average of three replicates and two layers in pond (surface and bottom).

of increases in TP and SRP were three and two times higher in T4 than in the two other treatments.

DISCUSSION

The importance of rabbits in integrated fish farming is to serve as a source of manure, an organic fertilizer that decomposes into water and releases nutrients. The lack of difference among the treatments tested in Rwanda

TABLE 3 Means ± standard errors for nutrients in water in tilapia ponds with three different rabbit stocking rates: 1, 2, and 4 per are. Samples were a mixture of water from the surface and bottom layers of the pond. Averages were calculated over the entire experimental period. Data that have same superscripts letters in the same row are not significantly different (P < 0.05)

Nutrients	Treatments		
	T1	T2	T4
1. Ammonia-nitrogen (mg.L^{-1})	0.44 ± 0.02^a	0.51 ± 0.18^a	0.69 ± 0.24b
2. Nitrite-nitrogen (mg.L^{-1})	0.40 ± 0.05^a	0.50 ± 0.07^a	0.69 ± 0.41b
3. Nitrate-nitrogen (mg.L^{-1})	0.75 ± 0.29^a	1.46 ± 0.25^a	2.72 ± 0.41b
4. Total nitrogen (mg.L^{-1})	13.52 ± 0.23^a	13.19 ± 0.88^a	12.45 ± 0.5a
5. Total phosphorus (mg.L^{-1})	0.62 ± 0.06^a	0.69 ± 0.07^a	0.94 ± 0.06b
6. Orthophosphate (mg.L^{-1})	0.07 ± 0.01^a	0.10 ± 0.01^a	0.21 ± 0.04b

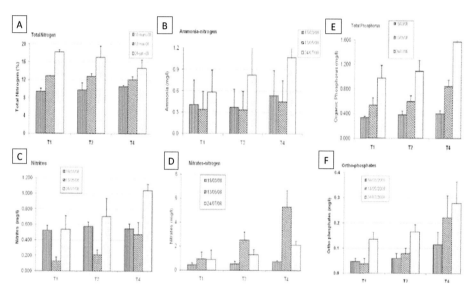

FIGURE 7 Nitrogen and phosphorus nutrient forms in tilapia pond water fertilized by rabbit droppings at three different rabbit densities on ponds (1, 2, and 4 rabbits per are) for 152 days.

suggests that the rabbit stocking rates used were quite low. Mean dissolved oxygen concentrations averaged around 5 mg/l in all treatments but the minimum recorded indicates a trend toward eutrophication (Mmochi et al. 2002). Lower DO concentrations were measured during morning hours indicating microbial activity (Mmochi et al. 2002; Tepe & Boyd 2002; Boyd & Queiroz 2001).

Mean nitrate-nitrogen in our study was relatively high (0.76, 1.46, and 2.72 mg/l for T1, T2, and T4, respectively) compared to 0.16, 0.12, 0.39 mg/l compared to ponds fertilized with cow dung mixed with compost, with green grass, and with compost, respectively (Rwangano 1990) and 0.12,

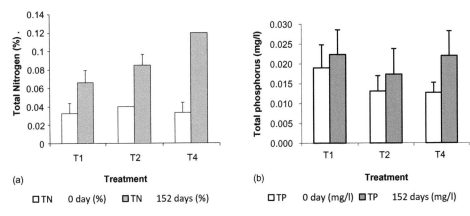

FIGURE 8 Total nitrogen (a) and total phosphorus (b) in sediments of fish ponds fertilised with 1, 2, and 4 rabbits per are at the beginning and end of a 152-day production period.

0.11 mg/l when treating ponds with ammonium sulfate and sodium nitrate (Tepe & Boyd 2002). Concentrations of soluble reactive phosphates were quite similar to phosphorus in ponds fertilized by green grass and chemical fertilizer (Tepe & Boyd 2002).

Nevertheless, primary productivity in our ponds was lower than other reported studies. The primary productivity in our ponds varied between 0.29 and 0.32 gC/m^2/d compared to 0.69–0.75 g C/m^2/d reported by Franco (1991) and 0.93 and 1.3 g C/m^2/d reported by Teichert-Coddington (1991) cited by Franco (1991). The observed trend of declining transparency should indicate an increase in primary production, but our results suggest that there may be another source of turbidity than phytoplankton as the pond water was most of the time brown (Boyd & Tucker 1998; Tepe & Boyd 2002). Alkalinity and pH were sufficient to have supported higher productivity (McNabb et al. 1989; Boyd 1979), further indicating that our rabbit dung loading rates were low.

Reflecting the low primary productivity, growth, and production of *O. niloticus* during this study was low, but within the range of results reported for tilapia growth in fertilized ponds with duck, chicken, and goat manure in China, Bangladesh, and India (Quazi & Huque 1991). Specific growth with rabbit droppings was slightly higher than that for fertilization with cow dung, grass, compost, and biodigestor effluent (Mahadevaswamy & Venkataraman 1988). Rabbit manure is more concentrated in phosphorus and nitrogen than other animal manure (Lebas et al. 1996), and its nitrogen is reported to be 10 times higher than that of cow dung (Kumar & Ayyapan 1998). The performance reported here can be considered as the lowest possible because the loading rate in rabbits didn't bring the ponds to a high level of primary production. More studies on other rabbit loading rates are needed to clearly establish the effect of different rates on the quality and management of fish ponds.

REFERENCES

Boyd, C. E, and J. F Queiroz. 2001. Nitrogen, Phosphorus loads vary by system. *The Advocate/Global Aquaculture Alliance USA* (December): 84–86.

Boyd, C. E. 1979. *Water quality in warmwater fish ponds*. Auburn, AL: Auburn University Agricultural Experiment Station.

Boyd, C. E., and C. S. Tucker. 1998. *Pond aquaculture water quality management*. New York: Kluwer Academic Publishers.

Breine, J. J., G. G. Teugels, and F. Ollevier. 1995. Résultats préliminaires de la pisciculture intégrée à la station de recherche piscicole de Foumban, Cameroun Paper read at l'Aménagement des écosystèmes agro-piscicoles d'eau douce en milieu tropical 16–19 mai 1995, at Bruxelles Belgique [Preliminary results of integrated fish farming at the fish breeding research station].

Breine, J. J., G. G. Teugels, N. Podoor, and F. Ollevier. 1996. First data on rabbit dung as a water fertilizer in tropical fish culture and its effect on the growth of *Oreochromis niloticus* (Teleostei, Cichlidae). *Hydrobiologia* 321(2): 101–107.

Dauta, A., and J. Capblancq. 1999. *Photosynthesis simulator PSS2000*. Toulouse, France: Centre d'Ecologie de Systèmes Aquatiques Continentaux (CESAC)/CNRS Université Paul Sabatier.

Descy, J. P. 1989. *Manuel pratique des techniques d'analyse de l'eau. Analyses physico-chimiques* [Practical manual of methods for water physico-chemical analysis]. Namur, Belgium: University Faculties Notre-Dame de la Paix, Namur.

Franco, L. 1991. *Nile Tilapia production in tropical microcosms fertilized with rabbit excreta*. Corvallis, OR: Oregon State University.

Franson, M. A. H., L. S. Clesceri, A. E. Greenberg, and A. D Eaton. 1985. *Standard methods for the examination of water and wastewater*. Washington, DC: American Public Health Association.

Kumar, K., and S. Ayyapan. 1998. *Integrated aquaculture in Eastern India, Working paper 5*. Purulia, India: Institute of Fresh Aquaculture.

Lebas, F., P. Coudert, H. Rochambeau, and R. G. Thébault. 1996. *Le lapin: Elevage et pathologie (nouvelle version revisitée)*, Vol. 19 [The Rabbit: Husbandry, Health and Production]. Edited by FAO.Rome: FAO.

Léonard, V., and C. K. Kanangire. 1998. *Recueil de Methodes d'analyses physico-chimiques de l'eau, des sols et des aliments* [Compilation of methods for water physico-chemical analysis, soil, and feeds]. Butare, Rwanda: Université Nationale du Rwanda/Facultés Universitaires Notre Dame de la Paix de Namur.

Mahadevaswamy, M., and L. V. Venkataraman. 1988. Integrated utilization of rabbit droppings for biogas and fish production. *Biolog. Wastes* 25(4): 249–256.

McCrosckey, R. 2001. Integration of rabbit production into populated areas, especially in hot climates. *Pan-Am. Rabbit Sci. Newsl.* 6(1): 18–20.

Micha, J. C. 2001. Systèmes d'élevages intégrés en zones humides tropicales Paper read at Aménagement des marais au Rwanda, 5–7 June 2001, at Butare-Rwanda.

Mmochi, A. J., A. M. Dubi, F. A, Mamboya, and A. W. Mwandya. 2002. Effects of fish culture on water wuality of an integrated mariculture pond system. *Western Indian Ocean J. Mar. Sci.* 1(1): 53–63.

Quazi, M., and Q. M. E. Huque. 1991. Integrated livestock-fish farming: Bangladesh perspective. Paper read at Aquaculture and Consumer Protection, 16–20 December 1991, Rome, Italy.

Rwangano, F. 1990. *Interactions of organic input types and water quality on the production of Oreochromis niloticus (Linnaeus, 1757) in Rwandan ponds.* Corvallis, OR: Oregon State University.

Tepe, Y., and C. E. Boyd. 2002. Nitrogen fertilization of golden shine ponds. *N. Am. J. Aquacult. Am. Fish. Soc. USA* 64:284–289.

Van Vleet, J. 1997. Raising rabbits over fish ponds. Paper read at the Third conference on the Culture of Tilapias at High Elevations in Africa, April 1997, at Rwasave Fish Farming Research Station, Rwanda.

Improving Water Use Efficiency in Semi-Arid Regions through Integrated Aquaculture/Agriculture

SAMI ABDUL-RAHMAN[1], I. PATRICK SAOUD[1],
MOHAMMED K. OWAIED[2], HANAFY HOLAIL[2], NADIM FARAJALLA[3],
MUSTAFA HAIDAR[3], and JOLY GHANAWI[1]

[1]*Department of Biology, Faculty of Arts and Sciences, American University of Beirut, Beirut, Lebanon*
[2]*Department of Biology, Faculty Science, Beirut Arab University, Beirut, Lebanon*
[3]*Faculty of Agriculture and Food Sciences, American University of Beirut, Beirut, Lebanon*

Two experiments were performed in the Bekaa plain in Lebanon to evaluate the feasibility of integrating aquaculture with established agriculture production in order to increase water productivity. Both experiments consisted of four plant management treatments: 1) Aquaculture effluent irrigation and no fertilizer; 2) aquaculture effluent irrigation and inorganic fertilizer; 3) well water irrigation and no fertilization; and 4) well water irrigation with inorganic fertilizer. In the first experiment, tilapia growth and radish production using aquaculture effluent were evaluated. All fish survived and grew, and radish production was improved by irrigating with aquaculture effluent. In the second experiment, maize (Zea mays) in large plots was irrigated with aquaculture effluent. Irrigation with effluent water improved maize production and improved soil nitrogen availability. In both experiments, fish production improved water value index and water use efficiency. Results suggest that aquaculture effluent can supplant inorganic fertilizers and could actually yield better crop production.

INTRODUCTION

During the next 50 years, problems associated with a lack of water will affect virtually everyone on the planet (Gleick 1992). At least one-third of the world's population lives in countries experiencing medium to high water stress (Stockholm Environment Institute (SEI) 1997). In the Middle East (ME), the water situation is precarious. Middle Eastern countries have experienced rapid population growth during the past few decades, while technical and institutional development in the water sector has been slow (Dillman 1989; Haddadin 1989; Gleick 1992; Arlosoroff 1995; Haddad & Mizyed 1996). The result has been reduced per capita fresh water availability. In the Mediterranean region, agriculture consumes about 72% of available freshwater resources (Hamdy & Lacirignola 1999) and thus will be the economic sector most affected by freshwater shortages (Water and Environmental Studies Center 1995). Lack of adequate supply is compounded by the fact that seasonal rains do not coincide with agricultural need (Hiessl & Plate 1990). Added to these is inefficient use of available water resources. Because irrigated agriculture uses more than 70% of available freshwater resources, any water-saving solution should start with improving irrigation water usage, mainly by implementing a "more crop per drop" approach (Seckler 1996). A possible method to increase crops per drop would be to integrate aquaculture into existing agriculture systems.

Integrated agriculture–aquaculture (IAA) allows for efficient use of water, particularly in arid and semi-arid regions (Hussain & Al-Jaloud 1998; Palada, Cole, & Crossman 1999; McIntosh & Fitzsimmons 2003). It potentially reduces cost of water and amount of fertilizer needed for crops and increases water productivity (Al-Jaloud et al. 1993; D'Silva & Maughan 1994, 1995; Azevedo 1998). Integrating agriculture with fish farming could improve fish pond water quality, reduce environmental impact of nutrient rich water discharge, reduce cost of water and amount of chemical fertilizer needed for crops, diversify farm production, increase income, and thus increase water efficiency (Billard & Servrin-Reyssac 1992; Ghate & Burtle 1993). If aquaculture is added to existing agriculture systems, it becomes a non-consumptive productive segment that does not compete with irrigation. Moreover, the long-term performance of diversified farms is better than non-diversified enterprises for various reasons. Just as with monocrop agriculture, monoculture fish farms operate at higher levels of risk from diseases, water quality, and price fluctuations (Naylor et al. 2000; Pant, Demaine, & Edwards 2005).

Various researchers have demonstrated the benefits of integrating agriculture with existing aquaculture facilities (Naegal 1977; Lewis et al. 1978; Watten & Buch 1984; Zweig 1986; McMurtry et al. 1990; Parker, Anouti, & Dickenson 1990; Olsen 1992; Al-Jaloud et al. 1993; McMurtry et al. 1993; Racocy, Hargreaves, & Bailey 1993; Seawright 1993; D'Silva & Maughan 1994, 1996; Palada, Cole, & Crossman 1999; Cruz et al. 2000; Al-Ahmed 2004). The present work evaluated the feasibility of increasing water productivity in semi-arid regions through integrated tilapia and crop production.

MATERIALS AND METHODS

The experiment was conducted during the summers of 2007 and 2008 at the Agricultural Research and Educational Center (AREC; 33°56' N, 36°05' E; 1000 m above sea level) of the American University of Beirut (AUB) in the North Central Bekaa plain of Lebanon. This being a semiarid region, almost all rainfall occurs during the cold and wet winter seasons (November–April), and summers are typically hot and dry in the daytime and cool at night. A weather station placed 200 m from the experimental unit indicated that average maximum temperature from July to September 2007 was 34.88°C, and average minimum temperature was 15.02°C. Relative humidity was 55.9%. Soil in the study area is alkaline (pH 8.0), clayey, vertic xerochrept formed from fine-textured alluvium derived from limestone (Ryan, Musharrafieh, & Barsumian 1980).

Experiment 1, 2007

AQUACULTURE

Six round 1-m^3 fiberglass tanks with central drains were stocked with size-sorted (17.9 ± 1.3 g; mean ± SE) Nile tilapia, *Oreochrormis niloticus*, fingerlings at 10 fish per tank. Fingerlings were spawned at the AUB aquaculture laboratory, and all came from a single brood. An air blower and submersible air diffusers provided continuous aeration and mixing of water. Fish were offered a commercial floating pelleted feed (35% crude protein; 5% lipid; Zeigler Bros. Inc. Gardner, PA, USA) at 5% body weight daily, divided into a morning (07:00–08:00) and afternoon (18:00–19:00) feedings. All fish from every tank were weighed biweekly to determine fish growth and to adjust ration. Every day about 10 liters of water from each tank were used for irrigation, and tanks were refilled with fresh well water to replace water lost due to irrigation and evaporation. Additionally, half of the water volume in each tank was replaced once a week. No filtration or other water treatment methods were used.

Water quality parameters were tested twice daily in the morning (07:00–08:00) and in the afternoon (18:00–19:00). Water temperature was measured using a pocket hygro-thermometer (EXTECH Instruments, London, United Kingdom). Dissolved oxygen (DO) concentrations and pH were determined using a portable pH and DO meter Model HQ-20 (Hach Company, Loveland, CO, USA). Total ammonia-nitrogen (TAN) and nitrite-nitrogen (NO_2-N) were determined twice a week using a fresh water test kit (Model AQ-2, La MOTTE Chemical Company, Washington, USA).

On termination of the experiment (after 60 days), fish in each tank were harvested and counted. All fish were group weighed and then weighed individually to the nearest 0.01 g. Specific growth rate (SGR) was calculated as: SGR = 100 × (ln W_f − ln W_i)/D, where W_f is the final mean weight of the fish, W_i is the initial mean weight, and D is number of days. Feed conversion ratio was calculated as FCR = F/(W_f − W_i), where F is the dry weight of feed offered to the fish, W_f is the final biomass of the fish, and W_i is biomass of fish at stocking.

AGRICULTURE

A plot of soil adjacent to the tanks was ploughed with moldboard twice then rototilled. Before planting, clods, rocks, and plant debris were removed. Twelve 1-m^2 plots were demarcated with flag markers. Seventy-six radish seeds (ASGROW Vegetables Seeds, Oxnard, USA) were planted per plot. Seeds were planted in four rows 100 cm in length and 20 cm apart, with seeds placed every 5 cm within each row. Radish was chosen as a test crop because radishes are easy to grow, have a short cultivation cycle of 30 to 40 days, and have a good market in the region where both the roots and the leaves are sold for human consumption. Biological insecticide (Bacill, Certis, USA) and selective insecticide (Marshal, FMC, USA) were applied at recommended doses. Plots were randomly divided into four treatments with three replicates per treatment. Treatments were: T1 = fish tank water irrigation with no additional fertilizer; T2 = fish tank water irrigation with fertilizer; T3 = well water irrigation with no additional fertilizer; and T4 = well water irrigation with fertilizer. Inorganic fertilizer NPK (15:15:15) at the rate of 100 kg/ha used commonly in the study area was applied to T2 and T4. Ten liters of fish effluent were applied every day at 6 pm to T1 and T2 using a sprinkling bucket, whereas T3 and T4 were irrigated using well water.

Plant growth was assessed by measuring plant height, new leaf number, and leaf length at 15 and 25 days after planting (DAP). At the end of the experiment, all plants in an area of 0.5 m^2 in the center of each plot were harvested. Average yield per plot was determined. Also fresh weight, total length, root length, tuber length, tuber diameter, leaf length, and leaf number were determined. Leaves and roots were dried at 70°C until constant weight and dry matter yield per treatment were determined.

A simple evaluation of water economics was performed. In a well-maintained system, water loss is mainly due to evaporation and fish biomass, which were negligible. Accordingly, consumptive water use (CWU) from the integrated system was due mainly to radish irrigation, which was calculated as m^3/m^2. Water use efficiency was calculated as: WUE kg/m^3 = total yield/CWU, and was used to evaluate efficiency for fish and radish integration. Finally, water productivity WP $/$m^3$ = WUE kg/m^3 × price $/kg, was calculated for fish and radish using an estimated fair market value for fish and radish.

Experiment 2, 2008

The second experiment was performed on a larger scale than the previous experiment to evaluate feasibility of an IAA system in reducing water use and increasing crop productivity on a small-holding farm scale. Fish growth was not evaluated. The crop used was 80-day maize to be used as silage.

AQUACULTURE

The five tanks used in the previous experiment were supplemented with a sixth tank that was filled with well water that was then used for irrigation with no fish or feed inputs (Figure 1). Two of the fish tanks were stocked with 60 fry (1.3 g) each, and two tanks were stocked with 10 male and 20 female brood stock tilapia each (biomass approximately 13.5 kg in each tank). The brood stock were offered 500 g of feed daily, and the fry were offered the same feed (ground to suitable size) to apparent satiation twice daily. Water quality parameters were measured daily in the morning (08:00) after irrigation and in the evening (20:00) as described above.

AGRICULTURE

A 500-m² field was tilled and prepared as described for experiment 1 and planted with rows of maize (*Zea mays*) with 70 cm between rows and 20 cm between seeds within each row. Half the field was fertilized at the locally common rate for NPK (17:17:17) of 100 kg/ha. Four 100-m² plots (5 m × 20 m) were delineated within the field, each containing eight rows

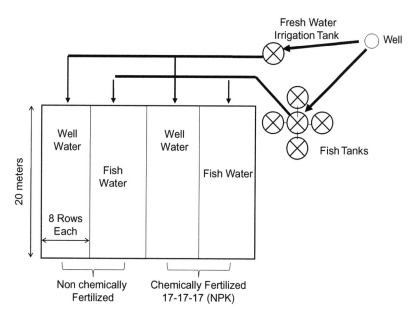

FIGURE 1 Schematic representation of the experimental setup used in Experiment 2.

of maize, with two plots in the fertilized section and two outside the fertilized section (Figure 1). Maize plants between rows were removed by hand after emergence. The whole field was irrigated with sprinklers and well water until emergence and then drip irrigation was fitted to every row. Treatments were: T1 = fish tank effluent/fertilized; T2 = well water/fertilized; T3 = fish tank effluent/unfertilized; T4 = well water/unfertilized. Fish tank water was pumped out of the sedimentation tank at 30 cm above bottom. The same irrigation pump was used for all treatments, and valves were used to separate well water from fish effluent. Each 100-m² plot was irrigated every second day with 1 m³ of water.

Plant growth parameters measured were plant height, plant weight, and number of leaves per plant. Experimental time was recorded as number of days after sowing (DAS). To achieve replications at sampling and harvest, four 1-m² quadrants were randomly delineated within each treatment plot, and plant height and number of leaves of all plants within the quadrant were measured weekly. At 54 DAS and every two weeks thereafter, all the plants within the quadrants were cut at ground level and aboveground biomass measured.

Soil samples were collected at a 15 to 30 cm depth prior to maize seeding, at maize seeding, and then every two weeks until harvest to measure nitrogen, phosphorous, and potassium concentrations. All soil samples were air-dried, ground, and sieved through a 1-mm mesh. A 5-g sample of air-dried soil was used to estimate soil P as described by Watanabe and Olsen (1965). Five-gram soil samples were treated with ammonium acetate (NH_4OAc) to extract potassium, after which K^+ concentrations were determined by flame photometry. Nitrogen content in the soil was determined using an elemental nitrogen analyzer (Thermo Finnigan/EA1112 elemental analyzer, Thermo Electron Corporation, Madison, WI, USA) with aspartic acid as a calibration standard.

Statistical Analysis

All statistical analyses were performed using SPSS statistical software (V.12 for Windows, SPSS Inc., Chicago, IL, USA). Means of parameters were estimated using one-way ANOVA and Student Newman-Keuls multiple-range test to determine significant differences ($P < 0.05$) among treatment means. Significant differences were considered when $P < 0.05$.

RESULTS

Experiment 1, 2007

AQUACULTURE

Fish survival was 100%, daily weight gain was 0.6 ± 0.02 g/fish/d (mean ± SE), average FCR was 2.78, and SGR was 1.9%/d. There were no differences in tilapia growth among the six culture tanks (Table 1), and

TABLE 1 Stocking Density, Initial Weight, Survival, Final Weight, Specific Growth Rate (SGR), and Feed Conversion Ratio (FCR) for Tilapia Reared in 1 m^3 Tanks in the Bekaa Plain During Summer Season. Data are Means (±SE) of all Tanks. No Significant Differences were Observed Among Tanks

Item	Average
Stocking Density/m^3	10 ± 0.0
Initial Weight/Fish (g)	17.8 ± 1.3
Survival (%)	100
Biomass at Harvest (kg/m^3)	1.11 ± 0.07
Final Weight (g/fish)	55.8 ± 3.1
Final Length (cm)	13.5 ± 0.4
SGR (%/day)	1.9 ± 0.09
Feed Offered (g)	1052.1 ± 0.0
FCR	2.78 ± 0.22

growth curves in all tanks were similar to each other throughout the study period (Figure 2). Water quality parameters remained within the safe limits for tilapia growth during the study. Maximum TAN concentration was 1.76 ± 0.26 mg/L (mean ± SE), and nitrite-N concentration was 0.61 ± 0.03 mg/L. Morning water temperature (at 7:00 h) averaged 19.67 ± 0.58°C, and evening water temperature (at 19:00 h) averaged 25.87 ± 0.53°C. Dissolved oxygen concentration (DO) ranged from 5 mg/L to 8.5 mg/L, and pH ranged from 7.6 to 8.6. Lowest values of dissolved oxygen, water temperature, and pH were routinely observed in early morning and high values in the evening. Effluent pH was 0.4 points greater than incoming well

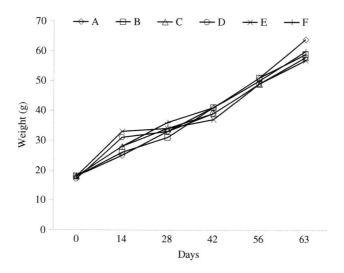

FIGURE 2 Growth performance of tilapia reared in 1 m^3 tanks in the Bekaa plain during summer season. Data points are means of 10 fish. Model average of all tanks is: y = 7.99x + 10.2; R^2 = 0.989.

water, and ammonia and nitrite nitrogen levels in well water were below detectable limits using the La Motte kit.

AGRICULTURE

Radishes in all treatments grew well and produced tubers. Growth parameters were not significantly different from each other among treatments at 15 and 25 days after planting (DAP), thus data are not reported nor discussed. At harvest (35 DAP), leaves in non-fertilized treatments irrigated with fish effluent (T1) were significantly larger than leaves in other treatments. Leaves in fertilized plots irrigated with well water (T4) and in fertilized plots irrigated with fish effluent (T2) showed no significant differences from each other. No significant differences in root length were observed among treatments. Both treatments irrigated with fish effluent produced tubers with significantly larger diameters than treatments irrigated with unfertilized well water. Plant length in fish effluent treatments was significantly larger than plant length in well water treatment (Table 2).

Significant differences in plant biomass and plant dry weight were observed among treatments. Leaf and tuber fresh weight in plots irrigated with fish effluent was significantly greater than those irrigated with well water. Non-fertilized plants irrigated with aquaculture effluent yielded greater leaf and tuber production than fertilized plants irrigated with well water. Tuber biomass of plants irrigated with fish effluent and treated with inorganic fertilizer (T2) were greater than tuber biomass of plants that received fish tank water only (T1), although not significantly different (Table 3).

In fish effluent treatments, total yield of radish plants was significantly greater than yield in both well water treatments. There were no significant differences in total dry and total fresh weights of plants between fertilized fish effluent treatments and non-fertilized fish effluent treatments.

TABLE 2 Tuber Length, Tuber Diameter, Root Length, Leaf Length, and Total Plant Length of Radish Grown in 1 m^2 Plots in the Bekaa Plain During Summer Season. Data are Means of 48 Radish Plants (±SE), and Values in the same Column Having Different Superscripts are Significantly Different ($P < 0.05$) from Each Other. Data were Collected 35 Days after Planting

Treatment	Tuber L (cm)	Tuber D (cm)	Root L (cm)	Leaf L (cm)	Total L (cm)
T1[1]	4.82 ± 0.1^a	3.83 ± 0.07^a	10.53 ± 0.44^a	14.96 ± 0.3^a	30.32 ± 0.53^a
T2[2]	4.73 ± 0.09^a	3.76 ± 0.07^a	11.36 ± 0.55^a	13.96 ± 0.31^b	30.07 ± 0.79^a
T3[3]	4.58 ± 0.15^{a,b}	3.48 ± 0.09^b	10.15 ± 0.42^a	12.57 ± 0.3^c	27.32 ± 0.62^b
T4[4]	4.35 ± 0.1^b	3.47 ± 0.06^b	9.96 ± 0.51^a	13.8 ± 0.27^b	28.12 ± 0.72^{a,b}
PSE[5]	0.12	0.10	0.26	0.20	0.3

[1]T1: Fish tank water irrigation and no fertilizer in plots.
[2]T2: Fish tank water irrigation with fertilizer in plots.
[3]T3: Well water irrigation and no fertilizer in plots.
[4]T4: Well water irrigation with fertilizer in plots.
[5]PSE: Pooled standard error.

TABLE 3 Tuber Fresh Weight, Leaf Fresh Weight, and Total Fresh Weight of Radish Grown in 1 m² Plots in the Bekaa Plain During Summer Season. Data are Means of 48 Radish Plants (±SE), and Values in the same Column Having Different Superscripts are Significantly Different ($P < 0.05$) from Each Other. Data were Collected at 35 DAP

Treatment	Tuber F Wt (g)	Leaf F Wt (g)	Total F Wt (g)
T1[1]	28.56 ± 0.65ᵃ	8.76 ± 0.14ᵃ	37.3 ± 0.78ᵃ
T2[2]	28.83 ± 0.15ᵃ	8.5 ± 0.02ᵃ	37.33 ± 0.14ᵃ
T3[3]	24.86 ± 1.14ᵇ	6.06 ± 0.2ᶜ	30.93 ± 1.32ᵇ
T4[4]	25.66 ± 0.2ᵇ	7.43 ± 0.07ᵇ	33.03 ± 0.27ᵇ
PSE[5]	0.30	0.13	0.33

[1]T1: Fish tank water irrigation and no fertilizer in plots.
[2]T2: Fish tank water irrigation with fertilizer in plots.
[3]T3: Well water irrigation and no fertilizer in plots.
[4]T4: Well water irrigation with fertilizer in plots.
[5]PSE: Pooled standard error.

Water productivity in IAA treatments (T1 and T2) was higher than in non-integrated treatments (T3 and T4). In integrated treatments, WVI was \$10.3/m³ for radish and \$1.5/m³ for fish, whereas in non-integrated or conventionally irrigated treatments (T4) WVI was \$9.11/m³ for radish alone. Consequently, radish-fish culture yields gross incomes that are 23% greater than radish monoculture (Table 4). WVI was increased by 12% for IAA treatments (T1 and T2) in comparison with well water treatment with NPK fertilization (T4) and by 17% in comparison with well water treatment without NPK fertilization (T3) after price of fertilizer is taken into account (Table 4). Water use efficiency based on plant biomass was 8.24 kg/m³ for aquaculture effluent treatments (T1 and T2), 7.29 kg/m³ for fertilized and well water-irrigated radishes (T4), and 6.83 kg/m³ for well water-irrigated but unfertilized radishes (T3) (Table 4).

TABLE 4 Water use Efficiency and Water Productivity for Radish Grown in 1 m² Plots and for Tilapia Reared in 1 m³ Tanks in the Bekaa Plain During Summer Season. The Price of Radish and Tilapia were Calculated at \$1.25/kg for Radish and \$7/kg for Tilapia

Treatment	WUE (kg/m³)	(WP \$/m³)
T1	8.24	10.3
T2	8.24	10.3
T3	6.83	8.53
T4	7.29	9.11
Tilapia	0.22	1.54

T1: Fish tank water irrigation and no fertilizer in plots.
T2: Fish tank water irrigation with fertilizer in plots.
T3: Well water irrigation and no fertilizer in plots.
T4: Well water irrigation with fertilizer in plots.

Experiment 2, 2008

AQUACULTURE

Tilapia fry survived and grew well. Two of the brood stock jumped out of their tank and were found dead and a third fish was not accounted for at harvest. All other fish survived and had no external signs of stress at harvest. The largest fish was 690 g, and the smallest fish was 210 g. Water temperature averaged 19.4 ± 0.02°C (mean ± SE) at 08:00, 18.4 ± 0.03°C (mean ± SE) immediately following irrigation and water replacement, and 27.5 ± 0.02°C (mean ± SE) at 20:00. Dissolved oxygen (DO) was 6.3 ± 0.05 mg/L at 08:00 and 5.2 ± 0.00 mg/L at 20:00. Total ammonia-N (TAN) and nitrite-N concentrations averaged 0.6 ± 0.1 mg/L and 0.4 ± 0.2 mg/L, respectively, throughout the experiment except from July 20, 2008, to July 23, 2008, when TAN levels increased above 3.0 ppm and nitrite-N levels were above 0.8 ppm. Water in the system was changed and water quality parameters then remained stable within acceptable limits for the remaining of the experiment.

AGRICULTURE

Maize plants fertilized and irrigated with fish tank effluent appeared taller with greater numbers of leaves than plants in all other treatments throughout most of the experimental period (Table 5; Figure 3A, B). At 20 DAS, there were no significant differences in maize plant height among treatments. During the rest of the experimental period, average maize plant height remained significantly greater in maize plants irrigated with fish effluent and fertilized (T1) than maize plant height in all other treatments (Figure 3A). The lowest plant height was observed in maize plants irrigated with well water and unfertilized. A similar trend was observed for number of leaves over the experimental period, where maize plants that were fertilized and irrigated with fish tank effluent had a greater number of leaves than all other treatments throughout the experimental period (Figure 3B). At harvest, weights of maize plants irrigated with fish effluent and unfertilized (T3) were significantly greater than the weights of maize plants in all other

TABLE 5 Growth Parameters (mean ± SE), Yield, and Water use Efficiency (WUE) of Maize Plants Irrigated with Different Sources of Water (Fish Effluent and Well Water and Fertilization)

Treatment	Plant height (cm)	Number of leaves per plant	Plant weight (g)	Yield (kg/m^2)	WUE (kg/m^3)	WVI ($/m^3)
T1	276.3 ± 0.73a	18.7 ± 0.16a	870.3 ± 2.30b	6.96b	1.91b	0.50b
T2	244.9 ± 0.71c	17.2 ± 0.14c	769.1 ± 2.22c	6.15c	1.69c	0.44c
T3	262.3 ± 0.86b	17.8 ± 0.12b	884.0 ± 2.91a	7.07a	1.94a	0.51a
T4	213.9 ± 1.32d	16.4 ± 0.11d	526.2 ± 3.25d	4.21d	1.16d	0.30d

T1 = drip irrigation with fish tank effluent/fertilized; T2 = drip irrigation with well water/fertilized; T3 = drip irrigation with fish tank effluent/unfertilized; T4 = drip irrigation with well water/unfertilized. Values in the same column sharing the same letter are not significantly different from each other ($\alpha = 0.05$).

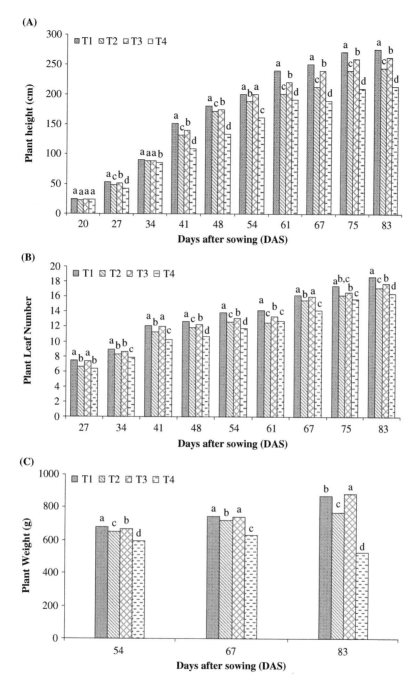

FIGURE 3 Growth parameters of maize plants irrigated with different sources of water (fish effluent and well water and fertilization): (A) height, (B) number of leaves, and (C) weight during experimental period of three months. T1 = drip irrigation with fish tank effluent/fertilized; T2 = drip irrigation with well water/fertilized; T3 = drip irrigation with fish tank effluent/unfertilized; and T4 = drip irrigation with well water/unfertilized. Data followed by the same letter do not differ significantly from each other ($\alpha = 0.05$).

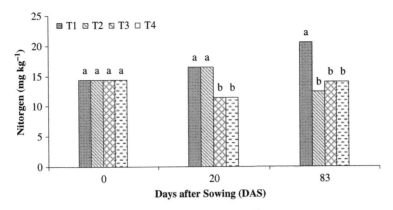

FIGURE 4 Effect of water sources for irrigation (fish effluent and well water) and fertilization (T1 = drip irrigation with fish tank effluent/fertilized; T2 = drip irrigation with well water/fertilized; T3 = drip irrigation with fish tank effluent/unfertilized; and T4 = drip irrigation with well water/unfertilized) on nitrogen (N) levels in soil samples. Data followed by the same letter do not differ significantly from each other ($\alpha = 0.05$).

treatments. Weights of maize plants irrigated with well water and unfertilized were significantly less than weights of maize plants in all other treatments (Table 5; Figure 3C).

Prior to crop planting and fertilizing, there were no significant differences in soil N, P, and K levels among plots. At the start of drip irrigation, (20 DAS), soil N in plots T1 and T2 was greater than soil N in T3 and T4 ($P < 0.05$). After the crop was harvested (83 DAS), the soil irrigated with fish effluent and fertilized (T1) had significantly greater levels of nitrogen than soils in all other treatments (T2, T3, and T4) ($P < 0.05$) (Figure 4). Mean soil phosphorus and potassium concentrations were above 20 mg/kg and 500mg/kg, respectively, in all treatments and thus cannot have affected plant growth results (FAO 2006).

Maize yield, water use efficiency, and water productivity data are shown in Table 5. Mean yield ranged from 4.21 kg/m² for T4 (well water and no fertilizer) to 7.07 kg/m² for T3, which was significantly greater than yield in all other treatments. The mean WUE and WVI of maize ranged between 1.16–1.94 kg/m²/m³ and 0.30–0.51 \$/m³, respectively. In T3, WUE (1.94 kg/m²/m³) and WVI (0.51\$/m³) were significantly greater than in all other treatments.

DISCUSSION

Experiment 1, 2007

AQUACULTURE

Growth performance and feed conversion efficiency were within normal limits for tilapia aquaculture. The FCR average of 2.78 obtained in the present study was similar to the average of 2.61 reported by Suresh and Lin (1992) for tilapia and was greater than the range of 1.45–2.40 reported by Yi & Diana (1996) for Nile tilapia in cages and also greater than 1.01–1.6 reported by Diana & Lin (2004) in ponds. FCR was probably high because constant change of water reduced primary productivity available to the fish. Moreover, daily temperature variations coupled with temperature changes during water exchange may have affected feeding behavior and growth rate, thus increasing FCR. Tilapia stocking density and production were low for aerated systems, but specific growth rates (1.8%/day) were good in comparison with other commercial scale aquaculture systems (McMurtry et al. 1990; Rakocy & Hargreaves 1993). The mean daily weight gain of 0.6 g/fish was lower than that obtained by Guerrero (1980) in ponds (0.56 to 1.60 g/fish) and lower than that obtained by Coche (1982) in cages (1.05 to 2.33 g/fish). However, fish in the reported studies benefited from natural productivity while fish in the present study did not.

Fish (55.86 ± 2.95 g; mean ± SE) did not reach market size. This was probably due to the relatively small stocking size (17.9 ± 1.3 g; mean ± SE) and short growout period (63 days). Balarin and Haller (1981, 1982) recommended stocking tilapia fingerlings (20 to 50 g) in commercial tanks and suggested a total production cycle duration of 205 days to produce market-size fish of 250 g average body weight. Such a growout period is not available in the Bekaa region, but fish stocked at 30 g have attained 160 g in 180 days (unpublished data). Moreover, individual fish growth is not only influenced by the initial size of the fish, but also by gender (Saoud et al. 2005), where all male tilapia fingerlings grow faster than females (Fitzsimmons 2000). Fish in the present study were not monosex, and size variability was apparent in all tanks. An all-male monosex culture would probably have yielded better average growth values. Nonetheless, the present study demonstrated that tilapia can be successfully grown in the Bekaa valley, and if cold tolerant strains, genetic variants such as GIFT tilapia, and/or greenhouses were used, harvest size could be attained in a shortened season.

Radish growth and biomass

Plants irrigated with aquaculture water grew better than those irrigated with well water. Radish tuber biomass, stem and leaf biomass, plant length, and root diameter (Tables 2 and 3) were higher when irrigated with aquaculture effluent compared to plants irrigated with well water. The low yields observed in the well water treatments suggest a lack of nutrients for adequate plant growth. The improved yields observed in the fish water treatments suggest that nutrients present in aquaculture effluent contributed significantly toward crop growth by meeting partial crop nutrient requirements. Results suggest that, even at low levels of nutrients present in aquaculture effluent, repeated applications yield better growth than a single large application of inorganic fertilizers at planting (Table 2). The fact that nutrients were available to plants in small amounts throughout the entire crop cycle allows for efficient on-time use by plants, increasing productivity (Pinto 1997). Results of the present experiment suggest that using aquaculture effluent for irrigation could reduce fertilizer use and increase economic returns. Differences in plant biomass were not significant ($P > 0.05$) among treatments treated with fish effluent, meaning that the use of fish effluents as irrigation and fertilizer source without application of inorganic fertilizer can produce crops equal in mass to crops treated with fish effluent and application of inorganic fertilizer (NPK). The lack of a significant difference between T1 and T2 can be attributed to constant availability of NPK in fish effluent, while fertilization only supplied NPK early and excess fertilizers might have leached into the ground due to repeated irrigation.

Water Value Index (WVI), Water Use Efficiency (WUE), and Profitability

In the IAA treatments, economic productivity of water and water value index were increased. Integrating aquaculture with agriculture increased water value index by 11% in comparison to the well water treatment with NPK fertilization. If we add the potential value of tilapia (1.5 $/m^3) (Table 4) and the reduction of fertilizers needed due to enrichment of water by aquaculture-generated metabolites, we find that an integrated system appreciably increases water value index. Our results were comparable with those reported by Hopkins et al. (1984), which indicated that integrating tilapia culture with agricultural crops would improve the economic viability of both crops by spreading the usefulness of the water systems. The returns on tilapia ($1.5 /m^3) were low and directly related to stocking density. A greater stocking density would provide greater revenue per integrated treatment as well as more nutrients in the water. However, higher stocking densities would also require more investment in feed and labor by the farmer whose primary source of income is assumed to be the vegetal crop with aquaculture used to increase returns and decrease CWU.

The present study shows that WUE of integrated treatments is greater than WUE of non-integrated treatments. Dey et al. (2004) showed that adoption of IAA technology increased water productivity substantially, improved total farm productivity by 10%, and increased farm income by 28% without additional consumptive use of water. However, we should remember that our calculations of water use do not take into account virtual water embedded in fish feed or other products needed for its production (see Allan 1998).

The amount of water used for fish production in the present study does not represent any additional demand on the existing water pumping rate on the farm. The water would be pumped on a daily basis even without fish production. The amount of fish produced in this system must be considered as an additional crop to other crops being produced. Therefore, aquaculture is a productive, low-consumptive segment of the water use system that does not compete with irrigation. In aquaculture, CWU is usually due to evaporation and leakage. If we assume that leakage is nil in a properly maintained system, then CWU is mainly due to evaporation and what gets incorporated into fish tissue. These elements of CWU equation are insignificant when compared to what gets lost during irrigation. Therefore, adding an aquaculture production unit between water source and agricultural end product increases productivity without significantly increasing CWU.

Experiment 2, 2008

Irrigation with tilapia aquaculture effluent without fertilization increased maize yield (7.07 kg/m^2), water use efficiency (WUE) (1.94 kg/m^3), and water productivity (0.51 \$/m^3) as compared to yield (6.15 kg/m^2), WUE (1.69 kg/m^3), and water productivity (0.44 \$/m^3) of maize irrigated with well water and fertilized at planting. The WUE of IAA irrigated maize was slightly greater than the 1.36–1.89 kg/m^3 reported by Katerji, Mastrorilii, & Rana (2008) for well water drip-irrigated maize in Lebanon. Furthermore, present results support findings of improved WUE using IAA for various other crops in other countries. Dey et al. (2010) reported that IAA farms in Malawi increased water productivity substantially, improved total farm production, and increased farm income without additional consumptive water use. Hussein and Al-Jaloud (1995) reported that irrigation with aquaculture effluents instead of well water significantly increased barley yields in Saudi Arabia.

The improvement in plant crop production in IAA is probably because of an increase in plant nutrients in the water (McIntosh & Fitzsimmons 2003). Seim, Boyd, and Diana (1997) reported a big increase in nitrogen and phosphorus content of aquaculture water over the course of a growout cycle. In the present study, phosphorus and potassium concentrations in the tank water remained very low throughout the study but that is because of constant water exchange when irrigating the maize. Ammonia nitrogen was constantly present in low concentrations. Because the soil in which the maize was planted contained more P and K than necessary for good maize growth (FAO 2006), we believe that growth differential among treatments was a result of continuous nitrogen input with every irrigation event. The increase in soil nitrogen in the treatments irrigated with aquaculture effluent supports such a statement. Nitrogen in the fish effluent must have partially supplied nutrient requirements for maize growth as evident in the improved yield and WUE of maize fertilized with fish effluent only. Apparently, availability of nutrients such as nitrogen in small amounts throughout the crop cycle allows more efficient nutrient absorption by the plants. Hussein (2009) observed that plant nutrient application through fertigation (mixing fertilizer into irrigation water) was much better used by maize as compared to conventional fertilization techniques.

CONCLUSION

As water stress increases in semi-arid non-industrialized nations, water use efficiency and food production will have to improve concurrently. One way of achieving this goal is to integrate aquaculture with agriculture. This should be done in a way that does not upset traditional lifestyles of local populations and increases economic returns to farmers. In the present work, integrating an aquaculture system between water source and traditional vegetable or grain agriculture increased productivity and economic returns without significantly influencing consumptive water use. The integration also decreased fertilizer needs and added animal protein production to the system without imposing a lifestyle change on the farmer. Further research on improving integration and extending fish growout seasons will greatly benefit traditional rural farmers.

REFERENCES

Al-Ahmed, A.A. 2004. Effluents and water quality in tilapia farms in Kuwait. In *Aquaculture–an ecologically sustainable and profitable venture. World Aquaculture Conference*, March 5, Honolulu, Hawaii, USA.

Al-Jaloud, A.A., G. Hussain, A. Alsadon, A.Q. Siddiqui, and A. Al-Najada. 1993. Use of aquaculture effluent as a supplemental source of nitrogen fertilizer to wheat crop. *Arid Soil Res. Rehab.* 7:233–241.

Allan, J.A. 1998. Virtual water: A strategic resource. *Ground Water* 36(4): 545–546.

Arlosoroff, S. 1995. Water resources management within regional cooperation in the Middle East. In *The Proc. of the second workshop on joint management of shared aquifers*, edited by M. Haddad, and E. Fietelson, held in Jerusalem, Israel, on Nov. 27–Dec. 1, 1994, 1–31.

Azevedo, C.M.S.B. 1998. Nitrogen transfer using ^{15}N as a tracer in an integrated aquaculture and agriculture system. PhD dissertation, University of Arizona, Tucson, Arizona.

Balarin, J.D., and R.D. Haller. 1981. Intensive tilapia culture: A scope for the future. *Proc: K.M.F.R.I. Symp. on aquatic resources in Kenya: A need for research*, 368–405.

Balarin, J.D., and R.D. Haller. 1982. *The intensive culture of tilapia in tanks, raceways and cages*. In *Recent advances in aquaculture*, edited by J.F. Muir, and R.J. Roberts, 267–355. London: Croom Helm.

Billard, R., and J. Servrin-Reyssac. 1992. Les impacts negatifs de la pisciculture d'e' tang sur l'environment [Negative impacts of pond pisciculture on the environment]. In *Production, environment and quality*, edited by G. Barnabe and P. Kestemont, 17–29. Oostende, Belgium: European Aquaculture Society

Coche, A.G. 1982. Cage culture of tilapias. In *The biology and culture of tilapias: ICLARM conference proceedings 7*, edited by R.S.V. Pullin and R.H. Lowe-McConnell, 205–243. Manila, Philippines: International Center for Living Aquatic Resource Management.

Cruz, E.M., A.A. Al-Ameeri, A.K. Al-Amed, and M.T. Ridha. 2000. Partial budget analysis of Nile Tilapia *Oreochromis niloticus* cultured within an existing agricultural farm in Kuwait. *Asian Fish. Sci.* 3:97–305.

Dey, M.M., P., Kambewa, D. Jamu, P. Paraguas, and M. Prein. 2004. *Impact of development and dissemination of integrated aquaculture-agriculture technologies in Malawi. Report to the Standing Panel on Impact Assessment (SPIA) of the Consultative Group of International Agricultural Research (CGIAR)*. Penang, Malaysia: The World Fish Center.

Dey, M.M., F. J. Paraguas, P. Kambewa, and D. E. Pemsl. 2010. The impact of integrated aquaculture–agriculture on small-scale farms in Southern Malawi. *Agric. Econ.* 41(1): 67–79.

Diana, J.S., and C.K. Lin. 2004. Stocking densities and fertilization regimes for Nile tilapia (*Oreochromis niloticus*) production in ponds with supplemental feeding. In *Proc. of the sixth international symposium on tilapia in aquaculture*, edited by R. Bolivar, G. Mair, and K. Fitzsimmons, 487–499. Manila, Philippines: BFAR.

Dillman, J. 1989. Water rights in the occupied territories. *J. Palestine Stud.* 19(1): 46–48.

D'Silva, A.M., and O.E. Maughan. 1994. Multiple use of water: integration of fish culture and tree growing. *Agrof. Syst.* 26:1–7.

D'Silva, A.M., and O.E. Maughan. 1995. Effect of density and water quality on red tilapia in pulsed flow culture systems. *J. Appl. Aquacult.* 5:69–75.

D'Silva, A.M., and O.E. Maughan. 1996. Optimum density of red tilapia *Oreochromis mossambicus* × *O. urolepis hornorum* in a pulsed-flow culture system. *J. World Aquacult. Soc.* 27:126–129.

FAO. 2006. *Near East fertilizer-use manual*. Cairo, Egypt: FAO, Regional Office for the Near East.

Fitzsimmons, K. 2000. Tilapia: The most important aquaculture species of the 21st century. In *Proc. of fifth international symposium on tilapia in aquaculture*, edited by K. Fitzsimmons and J.C. Filho, 3–8. Manila, Philippines: American Tilapia Association and ICLARM.

Ghate, S.R., and G.J Burtle. 1993. *Water quality in channel catfish ponds intermittently drained for irrigation*. In *Techniques for modern aquaculture*, edited by J.K Wang, 177–186. Joseph, MI: ASAE.

Gleick, P. 1992. Water and conflict. In *The workshop on environmental change, water resources, and international conflict*, held at the University of Toronto, June 1991.

Guerrero, R.D. 1980. Studies on the feeding of *Tilapia nilotica* in floating cages. *Aquacult.* 20:169–175.

Haddadin, M. 1989. Water resources in the Arab world and its strategic importance. In *Proc. of the conference on water resources in the Arab world, and its strategic importance*, held at Jordan University, Amman, April 2–4.

Haddad, M., and N. Mizyed. 1996. Water resources in the Middle East: Conflict and solutions. In *Proc. of the workshop on water peace and the Middle East: Negotiating resources of the Jordan River basin*, edited by T. Allan, 3–20. New York: Library Modern Middle East, Tauris Academic Studies.

Hamdy, A., and C. Lacirignola. 1999. *Mediterranean water resources: Major challenges towards the 21st century*. Bari, Italy: IAM Editions.

Hiessl, H., and E.J. Plate. 1990. A Heuristk closed loop controller for water distribution in complex irrigation systems. *Water Resour. Res.* 26(1): 1,323–1,333.

Hopkins, K.D., M.M. Hopkins, A. Al-Ameeri, and D. Leclerq. 1984. *Tilapia culture in Kuwait: A preliminary analysis of production systems. Kuwait Institute for Scientific Research, report no. 1252*. Kuwait: KISR.

Hussain, G., and A.A. Al-Jaloud. 1995. Effect of irrigation and nitrogen on water use efficiency of wheat in Saudi Arabia. *Agric. Water Manage.* 27:143–153.

Hussain, G., and A.A. Al-Jaloud.1998. Effect of irrigation and nitrogen on yield, yield components and water use efficiency of barley in Saudi Arabia. *Agric. Water Manage.* 36:55–70.

Hussein, A.H.A. 2009. Phosphorous use efficiency by two varieties of maize at different phosphorous fertilizer application rates. *Res. J. Appl. Sci.* 4(2): 85–93.

Katerji, N., M. Mastrorilii, and G. Rana. 2008. Water use efficiency of crops cultivated in the Mediterranean region: Review and analysis. *Eur. J. Agron.* 28: 493–507.

Lewis, W.M., J.H. Yoop, J. Schramm, and H.L. Brandesburg. 1978. Use of hydroponics to maintain quality of recirculated water in fish culture system. *Trans. Am. Fish. Soc.* 107:92–99.

McIntosh, D., and K. Fitzsimmons. 2003. Characterization of effluent from an inland, low salinity shrimp farm: what contribution could this water make if used for irrigation? *Aquacult. Eng.* 27:147–156.

McMurty, M.R., P.V. Nelson, D.C. Sanders, and L. Hodges. 1990. Sand culture of vegetables using recirculated aquaculture effluents. *J. Appl. Agricult. Res.* 5(4): 280–284.

McMurtry, M.R., D.C. Sanders, P.V. Nelson, A. Nash. 1993. Mineral nutrient concentration and uptake by tomato irrigated with recirculating aquaculture water as influenced by quantity of fish waste products supplied. *J. Plant Nutr.* 16:407–419.

Naegal, L.C.A. 1977. Combined production of fish and plants in recirculating water. *Aquacult.* 10:17–24.

Naylor, R. L., R.J. Goldburg, J.H. Primavera, N. Kautsky, M.C.M. Beveridge, J. Clay, C. Folke, J. Lubchenco, H. Mooney, and M. Troell. 2000. Effect of aquaculture on world fish supplies. *Nature* 405:1,017–1,024.

Olsen, G.L. 1992. The use of trout manure fertilizer for Idaho crops. In *National livestock, poultry and aquaculture waste management, ASAE Pub. 03–92*, edited by J. Blake, J. Donald, and W. Magette, 198–205. St. Joseph, MI: ASAE.

Palada, M. C., W. M. Cole, and S.M.A. Crossman. 1999. Influence of effluents from intensive aquaculture and sludge on growth and yield of bell peppers. *J. Sustainable Agric.* 14(4): 85–103.

Parker, D., A. Anouti, and G. Dickenson.1990. *Integrated fish/plant production system: experimental results. ERL report 90-34*. Tucson, AZ: University of Arizona.

Pant, J., H. Demaine, and P. Edwards. 2005. Bio-resource flow in integrated agriculture-aquaculture systems in a tropical monsoonal climate: A case study in northeast Thailand. *Agric. Sys.* 83(2): 203–219.

Pinto, J.M. 1997. Doses e períodos de aplicação de nitrogênio via água de irrigação na cultura do tomate [Doses and timing of application of nitrogen in the irrigation water of tomatoes]. *Horticultura Brasileira* 15(1): 15–18.

Rakocy, J.E., and J.A. Hargreaves. 1993. Integration of vegetable hydroponics with fish culture. A review. In *Techniques for modern aquaculture*, edited by J.K. Wang, 148–158. St. Joseph, MI: ASAE.

Rakocy, J.E., J.A. Hargreaves, and D.S. Bailey. 1993. Nutrient accumulation in a recirculating aquaculture system integrated with vegetable hydroponics. In *Techniques for modern aquaculture*, ASAE pub. 02–93, edited by J.K. Wang, 148–158. St. Joseph, MI: ASAE.

Ryan J., G. Musharrafieh, and A. Barsumian.1980. *Soil fertility characterization at the agricultural research and educational center of the American University of Beirut (AUB), publication no. 64*. Beirut, Lebanon: AUB.

Saoud I. P., D. A. Davis, L.A. Roy, and R. Phelps. 2005. Evaluating the benefits of size sorting tilapia fry before stocking. *J. Appl. Aquacult.* 17(4): 73–86.

Seawright, D.E. 1993. A method for investigating nutrient dynamics in integrated aquaculture-hydroponics system. In *Techniques for modern aquaculture*, ASAE pub. 02–93, edited by J.K. Wang, 137–147. St. Joseph, MI: ASAE.

Seim, W.K., C.E. Boyd, and J. S. Diana. 1997. Environmental considerations. In *Dynamics of pond aquaculture*, edited by H.S. Egna and C.E. Boyd, 163–182. Boca Raton, FL: CRC Press.

Seckler, D. 1996. *The new era of water resources management: "Dry" to "wet" water savings, Research report 1*. Colombo, Sri Lanka: International Water Management Institute (IWMI).

Stockholm Environment Institute (SEI).1997. *Comprehensive assessment of the freshwater resources of the world*. New York: Department for Policy Coordination and Sustainable Development, United Nations. Available online at: http://www.un.org/dpcsd

Suresh, A., and C. Lin. 1992. Tilapia culture in saline waters, a review. *Aquacult.* 106:201–226.

Watanabe, F.S., and S.R. Olsen.1965. Test of an ascorbic acid method for determining phosphorus in water and NaHCO3 extracts from soil. *Soil Sci. Soc. Am. Proc.* 29:677– 678.

Water and Environmental Studies Center (WESC). 1995. *Middle East regional study on water supply and demand development. Phase I study report submitted to the German Agency for Technical Cooperation (GTZ), the An-Najah National University, Nablus*. Nablus, Palestine: WESC.

Watten, B.J., and R.L. Buch.1984. Tropical production of tilapia (*Sarotherodon aurea*) and tomatoes (*Lycopersicon esculentum*) in small-scale recirculating water system. *Aquacult.* 41:271–283.

Yi, Y., C.K. Diana. 1996. Effects of stocking densities on growth of caged adult Nile tilapia (*Oreochromis niloticus*) and on yield of small Nile tilapia in open earthen ponds. *Aquacult.* 146:205–215.

Zweig, R.D. 1986. An integrated fish culture hydroponics vegetable production system *Aquacult. Mag.* (May/June): 34–40.

Growth Performance of Improved (EXCEL) and a Non-Improved Strains of *Oreochromis niloticus* Fry in a Recirculating Tank System in the UAE

NOWSHAD M. RASHEED and IBRAHIM E. H. BELAL

Aridland Agriculture Department, College of Food and Agriculture, United Arab Emirates University, Al Ain, United Arab Emirates

EXCEL is a new breed of O. niloticus developed by combining improved within family selection and rotational mating. A study was carried out to investigate early growth performance in an indoor recirculating aquaculture facility. The first study compared average body weight (ABW), average daily growth rate (ADGR), survival and food conversion efficiency (FCR) of first generation fry of imported EXCEL and non improved O. niloticus (NS) fry over an initial nursery stage from swim-up to 5.0 g (56 days, test 1) and a second nursery stage from ~3 to 15 g (28 days, test 2). ABW and ADGR of EXCEL fry was 4–5 times that of non improved tilapia during the first 56 days, which advantage was maintained over the second nursery stage also. In addition, FCR was lower in EXCEL compared with NS.

INTRODUCTION

Indoor Recirculating Aquaculture System (RAS) provide a higher degree of environmental control and require less water and less land area per kg of fish produced than most of the other production systems. They can be located in relatively close proximity to markets to reduce transportation cost, stress and mortalities during live transport. RAS can increase bio-security by minimizing the interaction of cultured fish with external biota to maintain integrity of both natural systems and the cultured stocks. (Watanabe et al. 2002).

Several fast-growing strains of tilapia have been produced for use in aquaculture. The GIFT or genetically improved farmed tilapia is a stock developed through combined family and within family selection (Eknath et al. 1993). The FaST is a strain developed through within family selection (Bolivar & Newkirk 2002). EXCEL, short for EXCELlent strain is a product of a selection program combining strain crosses and within family selection with rotational mating using the four parent lines, such as GIFT, FaST, Egypt strain and Kenya strain (Tayamen 2004).

The objective of the present research was to evaluate early growth performance, feed conversion ratio and survival of the first generation fry produced from EXCEL strain compared with a non-improved strain for use in an indoor recirculating aquaculture system.

MATERIALS AND METHODS

The experiments were conducted at the Indoor Aquaculture Recirculation Facility, College of Food and Agriculture, United Arab Emirates University, Al Ain, UAE. Two strains of *O. niloticus* were tested: the genetically improved EXCEL strain from the Philippines and a non-improved strain (NS) – offspring of an Egyptian strain from Ismalia.

Eight hundred (800) EXCEL fry (0.15 g average body weight) were imported in 2007 from the Bureau of Fisheries and Aquatic Resource – National Freshwater Fisheries Technology Center (BFAR-NFFTC), Philippine Department of Agriculture. The NS was the offspring of a pure stock of the Egyptian Ismalia strain imported (two thousand numbers) from the Kingdom of Saudi Arabia in 2005. The introduced stock was grown in fiberglass tanks (each 1 m^3) and spawned to obtain enough fry to conduct the experiments. The experiments described here were carried out on first generation fry of the introduced fish.

All fish were cultured in fiberglass tanks provided with central drainage pipes surrounded by outer sleeves pipes, perforated at the bottom, to facilitate self-cleaning. Settling tanks removed suspended and settled particles, a head tank maintained water pressure and a UV tube was included to remove germs. Heaters with thermostats maintained a constant water temperature of 28°C. The culture system was provided with a biological filter (plastic tubing structures), for controlling toxic inorganic compounds such as ammonia and nitrites. All tanks were continuously aerated with an air compressor (Hick Hargreaves U.K). Settled solids were removed manually by siphoning every day. Daily up to 5% of the total water volume was exchanged. Three tanks were assigned for each treatment. For rearing fry and fingerlings, the systems used were comprised of fiber tanks with effective volume of 38 liters (0.038 m^3) and 60 liters (0.06 m^3) respectively.

Dissolved oxygen, and pH were monitored daily and ammonia, nitrite and nitrate were monitored twice in a week. Dissolved oxygen and pH were measured using an YSI model 58 oxygen meter, and a Jenway 3205 pH meter, respectively. Ammonia, nitrite and nitrate were monitored using an Aquafast II (Orion, Germany) colorimeter.

The feed used for this study (for fry and fingerlings) was prepared in the laboratory of the UAE University and was formulated as recommended by (Uchida & King 1962, Santiago et al. 1985, Wee & Tuan 1988) using fish meal (50%), vegetable oil (3%), vitamins and mineral mixes (2%), and wheat powder, with proximate composition of 37.4% crude protein, 10.2% crude fat, 4.1% crude fiber, 8.6% moisture, 12.7% total ash and 190 (kJg-1) total energy.

Free swimming fry were fed at 25% of total fish body weight (ABW) for the first 2 weeks, followed by 20% of ABW for the next 2 weeks. The feeding rate was then reduced to 15% per day for another two weeks and finally reduced to 10% for the rest of the experiment. The feeding rate for fingerlings (3 g onwards) was 10% of total ABW per day for the initial 2 weeks and was then reduced to 8% per day for the rest of the experiment. Feeding frequency scheduled was 7 times with an interval of 1–1.5 hrs for fry and 4 times with an interval of 3–4 hrs for fingerlings, six days in a week. The fish were hand-fed with powdered feed for fry and with crumbles/pellets (0.3 mm to 1 mm) for fingerlings. The total fish weight in each tank was measured weekly to adjust the amount of feeding.

Experimental Design

Two consecutive tests were conducted to compare growth rate, feed conversion ratio and survival of the improved (EXCEL) strain and the non-improved strain (NS). Test 1 compared performance during a first nursery stage from swim-up to 5 g. As described above, first generation fry (ABW 0.01 g) from imported EXCEL brood fish and similarly sized NS brood stock were each stocked into triplicate 38 liter fiberglass tanks at 2 L^{-1} (2000 m^{-3}). The flow rate was adjusted to approximately 1 L min^{-1}. The duration of study was 56 days.

The second test was conducted to compare growth performance of both strains in the second nursery stage from 3 g to 15 g. EXCEL and NS fingerlings of 2.98 and 2.88 g, respectively, were stocked into triplicate 60 liter at 0.7 L^{-1} (700 m^{-3}). The flow rate was approximately 1.25L min^{-1} and duration of study was 28 days.

Every week and at the end of the studies, all the fish in each tank were counted and weighted and average wet body weight (ABW), average daily growth rate (ADGR), food conversion ratio (FCR) and percentage of survival of fry and fingerlings were calculated as below:

ABW = (Weight of total fish (g) sampled/Number fish sampled)

ADGR = (Total weight gain per fish (g)/culture days)

FCR = (total weight of dry feed (g)/total wet weight fish biomass gained (g))

% Survival = 100 (final total fish number/initial total fish number)

At the end of experiment, means of body weight (ABW), daily growth rate (ADGR), feed conversion ratio (FCR) and survival rate were subjected to one-way analysis of variance at $\alpha = 0.05$.

RESULTS

Growth performance in both tests is summarized in Table 1a, b. In the first test, the EXCEL strain ended with significantly higher mean body weight and daily growth rate (Table 1a) relative to the non-improved NS strain. The EXCEL fry averaged 5.33 g in 56 days at a stocking density of 2 L^{-1} while NS attained an average body weight of 1.53 g (Table 1a). In the second test,

TABLE 1 Growth Performance of EXCEL and Non-Improved Tilapia Fry

	EXCEL	Non-improved (NS)
a) TEST 1		
1. *Stocking Data*		
Number/tank	76	76
Stocking Density (L^{-1})	2	2
ABW (g)	0.01 ± 0	0.01 ± 0
2. *Harvest Data*		
Culture period (days)	56	56
ABW (g)	5.33 ± 0.33	1.53 ± 0.14
ADWG (g fish^{-1} day^{-1})	0.091 ± 0.003	0.026 ± 0.002
FCR	0.98 ± 0.05	1.28 ± 0.07
Survival (%)	95.17 ± 0.76	92.54 ± 2.73
b) TEST 2		
1. *Stocking Data*		
Number/tank	42	42
Stocking Density (L^{-1})	0.7	0.7
ABW (g)	2.98 ± 0.16	2.88 ± 0.102
2. *Harvest Data*		
Culture period (days)	28	28
ABW (g)	14.51 ± 0.42	8.33 ± 0.15
ADWG (g fish^{-1} day^{-1})	0.411 ± 0.02	0.194 ± 0.005
FCR	1.20 ± 0.04	1.48 ± 0.10
Survival (%)	92.85 ± 2.38	91.26 ± 2.74

TABLE 2 Average Water Quality Parameters

Parameter	Fry rearing system	Fingerling rearing system
Temperature (°C)	28.0	28.0
DO (mg L^{-1})	9.5	9.2
pH	7.2	7.5
Unionized Ammonia (mg L^{-1})	0.075	0.05
Nitrite (mg L^{-1})	1	3.0
Nitrate (mg L^{-1})	5	10

EXCEL fingerlings with ABW 2.98 g at stocking density of 0.7 L^{-1} reached 14.51 g in 28 days while NS with initial ABW of 2.88 g reached only 8.33 gd (Table 1b).

The ADGR as was also higher for EXCEL compared to NS. The ADGR of EXCEL was 0.091 g fish^{-1} day^{-1} and 0.41 g fish^{-1} day^{-1} in the first and second tests respectively. For the NS, ADGR was lower, 0.027 g fish^{-1} day^{-1} and 0.19 g fish^{-1} day^{-1} in the first and second tests respectively.

The FCR was significantly lower ($P < 0.05$) in EXCEL (0.98 and 1.2 in the first and second test respectively) than NS (1.28 and 1.48 in the first and second test respectively). Survival rate, 95.17% and 92.85% in the first and second test respectively for EXCEL and 92.54% and 91.26% in first and second test respectively for NS, was not significantly different ($P < 0.05$) among strains. Most of the mortality occurred shortly after the stocking and no mortality was observed during the progress of the tests in both strains.

Average water quality parameters are presented in Table 2. In all tanks and treatments, water quality parameters were similar and were within the safe limits.

DISCUSSION

Though the EXCEL out-performed NS in both tests, the growth performance of EXCEL fry was 4–5 times higher than NS in the first nursery stage and only double the NS in the second nursery stage. Even so, the significant shortening of the production cycle facilitated by the EXCEL would be beneficial to the hatchery operators to increase their production cycles per season.

There are no reports in the literature on the growth performance of the EXCEL strain in recirculating systems. However, similar experiments in other improved strains of Nile Tilapia have been reported. Ridha (2006) in Kuwait observed significantly higher growth performance of improved GIFT and SL strains compared to a non-improved strain from swim up to 1.0 g with daily growth rates of 0.021 and 0.033 g fish^{-1} day^{-1} in GIFT and SL respectively, compared to 0.018 g fish^{-1} day^{-1} in the non-improved strain, this at a stocking density of 1.65 fry L^{-1} whereas daily growth rate obtained in the present study was 0.091 g fish^{-1} day^{-1} and 0.026 g fish^{-1} day^{-1} for EXCEL and non improved strains respectively at a higher stocking density of 2 fry L^{-1}. Even with higher protein food (40% CP), El Sayed (2002) achieved daily growth rates of 0.021 g fish^{-1} day^{-1} over 40 days (initial average body weight 0.016 g) with the same non-improved *O. niloticus* used in the present study, even at a higher stocking density of 3 fry L^{-1} is comparable with the growth rate obtained for NS in this study.

In the second nursery stage, the significantly higher values obtained in EXCEL strain; suggest that this improved strain would reach a size of 15 g, about half the time required for the non improved strain. This would reduce the period of production cycle. Ridha (2006) reported that improved GIFT and SL fry with an initial size of 1 g fed with a 50% protein diet over 42 days achieved a daily growth rate of 0.38 and 0.37 g fish^{-1} day^{-1}, respectively, compared to 0.24 g fish^{-1} day^{-1} in a non-improved strain, results that are comparable with the growth rate obtained in this study.

Mamun et al. (2004) observed comparatively higher growth rates and lower feed utilization efficiencies in different strains of Nile tilapia (sex-reversed GIFT, mixed sex GIFT and mixed sex CNT) and speculated that the better growth performance of the improved strain GIFT tilapia might be due to behavioral factors rather than actual physiological growth potential because they could not find any significant differences in the metabolic efficiency between the improved and non-improved strains of Nile tilapia. However, FCR obtained in the first nursery stage (fed with approximately 35% protein) in the EXCEL strain (0.98) was significantly lower than in the NS group (1.28) and in the second nursery stage also FCR value was lower in EXCEL (1.2) than NS (1.48). Ridha (2006), using a marine diet containing 50% CP achieved even better values for FCR in improved strains (0.80 and 0.83) compared to a non-improved strain (0.89) in initial nursery stage. In the second nursery stage, results were similar with FCR of 0.96 and 1.06 in GIFT and SL, respectively, compared to 1.04 in the non-improved strain).

These results for initial growth performance of EXCEL show that this strain is excellent in a recirculation tank system and FCR and survival are favorable. However, further investigations are required to determine the growth performance of EXCEL strain to the market size (>200 g).

ACKNOWLEDGEMENTS

The authors would like to thank Mr. Mamdouh Kawanna for his assistance and Mr. Al Howaity for running the statistical analysis of the results.

REFERENCES

Bolivar, R.B., Newkirk, G.F., 2002. Response to within family selection for body weight in Nile Tilapia (*Oreochromis niloticus*) using a single-trait model. Aquaculture 204, 371–381.

Eknath, A.E., Tayamen, M.M., Palada-de Vera, M.S. Danting, J.C. Reyes, R.A. Dionisio, E.E. Capili, J.B. Bolivar, H.L Abella, T.A. Circa, A.C. Bentsen, H.B. Gjerde, B., Gjedrem, T., Pullin, R.S.V., 1993. "Genetic improvement of farmed tilapia: the growth performance of eight strains of *Oreochromis niloticus* tested in different farm environment". Aquaculture 111, 171–188.

El-Sayed, A.-F.M., 2002. Effects of stocking density and feeding levels on growth and feed efficiency of Nile tilapia (*Oreochromis niloticus* L.) fry. Aquac. Res. 33, 621–626.

Mamun S.M., Focken U., Francis G., Becker K. 2004. Growth performance and metabolic rates of genetically improved and conventional strains of Nile tilapia, Oreochromis niloticus (L.), reared individually and fed ad libitum. In: Proceedings of the Sixth International Symposium on Tilapia in Aquaculture, Manila, Philippines (ed. by R.Bolivar, G. Mair & K. Fitzsimmons), pp. 379–399. BFAR, Philippines.

Ridha, M.T., 2006. A comparative study on the growth, feed conversion and production of fry of improved and non improved strains of the Nile tilapia, *Oreochromis niloticus*. Asian Fish. Sci. 19, 319–329.

Santiago C.B., Aldaba M.B., Abuan E.F., Laron M.A. 1985. The effects of artificial diets on fry production and growth of Oreochromis niloticus breeders. Aquaculture 47, 193–203.

Tayamen, M. M., 2004. Nationwide Dissemination of GET-EXCEL tilapia in the Philippines. In: Proceedings of the Sixth International Symposium on Tilapia in Aquaculture, Manila, Philippines (ed. by R.Bolivar, G. Mair & K. Fitzsimmons), pp. 74–88. BFAR, Philippines.

Tayamen, M. M., Abella, T. A., 2004. Role of Public Sector in Dissemination of Tilapia Genetic Research Outputs and Links with Private Sector. Paper presented during the Workshop on Public-Private Partnership in Tilapia Genetics and Dissemination of Research Outputs. Days Inn Hotel, Tagaytay City, Philippines, 21–23 January 2004.

Uchida N.R., King J.E. 1962. Tank culture of tilapia. U.S. Fisheries and Wildlife Services. Fish. Bull. 62, 21–52.

Watanabe, W.O., T. M. Losordo, K., Fitzsimmons F. Hanley., 2002. Tilapia production systems in the Americas: technological advances, trends and challenges. Rev. Fish. Sci. 10: 465–498.

Wee K.L., Tuan N.A., 1988. Effects of dietary protein level on growth and reproduction in Nile tilapia (*Oreochromis niloticus*). In: Pullin, R.S.V., Bhukasawan, T., Tonguthal K., Maclean J.L. (Eds.) the 2nd Int. Symp. On Tilapia in Aquaculture. Bangkok, Thailand. ICLARM, Manila, Philippines. pp. 401–410.

Use of Underground Brackish Water for Reproduction and Larviculture of Rainbow Trout, *Oncorhynchus mykiss*

M. MOHAMMADI[1], H. SARSANGI[1], M. ASKARI[2], A. BITARAF[3], N. MASHAII[3], F. RAJABIPOUR[1], and M. ALIZADEH[3]

[1]*Researcher of Iranian Fisheries Research Organization, Bafgh Inland Saline Water Fish Research Station, Yazd, Iran*
[2]*Department of Biology, Faculty of Science, Shahid Bahonar University of Kerman, Iran*
[3]*Member of Scientific Board of Iranian Fisheries Research Organization. Bafgh Inland Saline Water Fish Research Station, Bafgh, Yazd, Iran*

Broodfish were reared for 6 months in underground brackish water (11 ppt), and the quality of eggs and semen was evaluated in comparison to broodfish held in fresh water. Fry produced by the broodfish in brackish water (S) and freshwater (F) were then reared in different salinities (1, 4, 7, and 10 ppt). Broodfish successfully matured and spawned in underground brackish water, and gonad quality was better than of the broodfish in fresh water ($p < 0.01$). Growth indices of fry produced by both groups decreased with increasing salinity ($p < 0.01$). Survival was not affected by the salinities tested. Underground brackish water can be used as a source to induce successful sexual maturation and to produce high-quality gonads.

This research was financed by the Iranian Fisheries Research Organization. We thank our colleagues at Bafgh Inland Saline Water Fish Research Station (BRS) and at Sahid Motahhari Cold Water Fish Genetic and Eugenics Research Center (SMC), especially Dr. Gorgipour, H. Gandomkar, R. Ansari, and H. Moradian for their help.

INTRODUCTION

The increase of saline groundwater has been become a major agricultural problem in many regions. This water has been used for culture of marine and euryhaline finfish (Primary Industries and Resources South Australia 1999). Fielder, Bardsley, and Allan (2001); Ingram, McKinnon, and Gooley (2002); and Partridge and Lymbery (2008) used saline groundwater in their experiments on Australian snapper, *Pagrus auratus*, selected aquatic animals, and barramundi, *Lates calcarifer*, respectively. There were some attempts to evaluate the effects of brackish water on fish maturation, reproduction, growth, and survival (Albrektsen & Torrissen 1988; Altinok & Grizzle 2001; Handeland et al. 2003, etc.).

Even for fish normally found only in freshwater, brackish groundwater might be useful. Nafisi (2002), for example, documented higher growth rate for rainbow trout (*Oncorhynchus mykiss*) in underground brackish water and faster gonad maturation than in freshwater (Falahati 2003). This study was designed to evaluate the effects of brackish groundwater on broodstock rearing, gamete quality, and larviculture of rainbow trout.

MATERIALS AND METHODS

Broodstock Rearing & Gamete Quality

One hundred 3-year-old broodfish ($♂/♀ = 1/3$) with an initial body weight of 1379.9 ± 292.84 g were randomly selected in April 2007 and stocked in freshwater in a flow-through concrete pond (flow rate = 1 cm/s) in Shahid Motahhari Cold Water Fish Genetic and Eugenics Research Center (SMC; 30°43´N, 51°34´E; altitude: 1860 m; yasouj) in. Six months before reproduction, 33 of these broodfish ($♂/♀ = 1/3$) were transferred to the Bafgh Inland Saline Water Fish Research Station (BRS; 31°37´N, 55°17´ E; altitude: 990 m) in August 2007 and, after a 24-h acclimation, were stocked into three aerated 3 m^3 fiberglass tanks with $♂/♀ = 3/7$, 6 kg/m^3 density, and 10-17 l/min water flow. Fish were fed 1% of body weight (twice per day) with commercial feed pellets (CP 43%, CL 14%, CF 4%, P 0.8%, and moisture 10%). Broodfish were reared and matured under similar conditions in brackish water (11 ppt) at BRS (treatment) and SMC (control) with freshwater for 6 months.

Every month, fish were anesthetized with 150–200 ppm clove meal *Eugenia caryophyllata* (Nafisi & Falahati 2008), measured for standard length and live body weight. After six months, fish were deemed mature through visual examination and stripped. Egg diameter was evaluated as an indicator of egg quality, and the total extracted eggs were weighed without ovarian fluid. Absolute and relative fecundity was calculated based on the number of all extracted eggs and number per female body weight, respectively. Immediately after stripping, weight and diameter of 30 eggs were measured by using a digital balance (EK 600i; 0.01 g accuracy) and calipers (0.1 mm accuracy), respectively. Sperm motility of broodfish for 5 males was estimated by microscopic observation (Ciereszko & Dabrowski 1995) four times per fish. Sperm concentration was also determined four times per fish with a double Neubauer counting chamber, HBG (Poole & Dillane 1998).

Larviculture

Fertilization, incubation, and hatching were conducted at both sites under standard conditions (in freshwater). Fry originating from freshwater broodfish and reared in freshwater at SMC (F) were then transferred to the BRS and acclimatized in dechlorinated fresh tap water for 10 days. Fry reared in freshwater in BRS and originating from broodfish kept in underground brackish water (S) and F fry of initial body average weight 0.3 ± 0.05 g were then randomly stocked at the rate of 3.3 fish per liter into 24, 15-L aquariums. Tanks were aerated and water flow maintained at 0.6 l/min. Following 3 days of acclimation, water salinity in the aquaria was adjusted with a mixture of dechlinated tap water and underground brackish water to salinity levels of either 1, 4, 7, or 10 ppt, for a total of 8 treatments each with 3 replicates. Salinity levels were adjusted to the nearest 0.09 ppt. Fish were fed to satiation with commercial feed pellets (50% CP, 18% CL, 10% ash, 1.3% CF, and 1.6% total phosphorous) four times per day (8:00, 12:00, 16:00, and 20:00 hours) for 6 weeks. Each week, fry were anesthetized in 150 ppm clove meal, and the total length and weight were measured as for the eggs, above. Specific growth rate (SGR), body weight increase (BWI), food conversion rate (FCR), and weight gain (WG) were used as growth indices as follows:

$$SGR = \{(LnW_f - LnW_i)/T\}*100$$
$$BWI = (W_f - W_i)/W_i*100 \ (\%)$$
$$FCR = F_i/(W_f - W_i)$$
$$WG = W_f - W_i \ (g)$$

where W_f = final weight (g); W_i = initial weight (g); F_i = food intake (g); and T = rearing duration.

In all experiments and treatments, water temperature, dissolved oxygen, oxygen saturated percentage, pH, salinity, and electrical conductivity (Ec) were measured with digital WTW-330. De-ionic ammonium and nitrite were measured with a Perkin-Elmer spectrophotometer (lambda 25) based on the phenate and colorimetric methods, respectively (Clesceri, Greenberg, & Eaton 1998). The heavy metals copper, iron, nickel, and zinc were measured by atomic absorption (PerkinElmer 3110). Calcium, chloride, magnesium, alkalinity, hardness, sulfate, and total dissolved solid (TDS) were measured by titration methods in the Ab-o-khak laboratory at the Yazd Natural Science Research Center.

Data were analyzed by SPSS (version 11.5). Brood stock condition and gamete quality in fresh and brackish water were compared with independent sample t-tests. Fry growth indices of were analyzed by one-way ANOVA and Duncan's new multiple range test. Interaction between different salinity levels and broodfish rearing in either fresh or brackish water was assessed with a covariance test.

RESULTS

Water quality parameters are given in Table 1. There were no significant differences in temperature (20.3 ± 1.06°C), dissolved oxygen (6.1 ± 0.5 ppm), pH (8.26 ± 0.09), ammonia (0.018 ± 0.003 ppm), or nitrite (0.094 ± 0.061 ppm) in all experimental groups (p > 0.01).

No significant differences in weight, length, and condition factor of brackish and freshwater broodstocks were observed (p > 0.01). Likewise, no significant differences were observed in relative fecundity and sperm motility in brackish and freshwater broodstocks. However, egg diameter and

TABLE 1 Physicochemical Factors of Fresh and Brackish Water

Factors	Freshwater	Brackish water
Temperature (°C)	10–12	10–12
Ec (mmos/cm)	0.354	19.21–22.7
Salinity (ppt)	0	9.1–13.3
NO_2 (ppm)	0.0231	0.06
Calcium (ppm)	102.4	328
Magnesium (ppm)	31	150
Sodium (ppm)	–	3565*
Potassium (ppm)	–	37.6*
Cupper (ppm)	<0.001	<0.001
Iron (ppm)	<0.001	<0.001
Nickel (ppm)	–	<0.001
Zink (ppm)	–	<0.001
Dissolve Oxygen (ppm)	>7	>5.5
pH	7.24	8.13
NH_3 (ppm)	0.0015	0.002
Free chloride (ppm)	<0.001	–
Chloride (ppm)	460.85	4042*
Bicarbonate (ppm)	97.65	93
Sulfate (ppm)	283.2	3456
Alkalinity (ppm)	170	125
Hardness (ppm)	386.2	1450
TDS (ppm)	1008.6	12270

Sourced from Mashaii (2006).

TABLE 2 Relative Fecundity and Gonad Quality of Broodstocks in Fresh and Brackish Water

Items	Treatments	
	Brackish	Fresh
Relative fecundity (eggs/kg female)	1355.74 ± 525.94[a]	1253 ± 306.86[a]
Egg diameter (mm)	5.4421 ± 0.20[b]	5.19 ± 0.38[a]
Sperm motility (seconds)	42.63 ± 1.17[a]	39.34 ± 1.23[a]
Sperm concentration (sperm/ml)	$26*10^9 \pm 1.41*10^9$ [b]	$18*10^9 \pm 1.31*10^9$ [a]

Numbers with same letters are not significantly different ($p > 0.01$).

sperm concentration in brackish water were significantly higher compared to broodstock maintained in freshwater (Table 2).

At all salinity levels, BWI and SGR were slightly higher ($p > 0.01$) in fry whose parents were held in brackish water (S group). There was no significant interaction between broodstock groups and salinity level on the growth indices of the fry ($p > 0.01$). In both F and S groups, WG, BWI, and SGR significantly decreased with increased salinity ($p < 0.01$) (Table 3). There was no difference in FCR among all salinity levels for both groups and also between the two groups at each salinity ($p > 0.01$). More than 96% survival was observed in all treatments with no significant differences among groups ($p > 0.01$).

DISCUSSION

We observed successful gonad development and spawning in rainbow trout held in brackish water, and others have supported our observations. Fast et al. (1991), for example, showed that Pacific salmon, *Oncorhynchus sp.*, held in marine water (34 ppt) matured and spawned successfully. Albrektsen and Torrissen (1988) reported that rainbow trout and their eggs (until the eyed stage) had better survival in brackish water than fresh, especially at higher temperatures. Atse (1999) found that gonad development of both male and female Arctic char *Salvelinus alpinus* broodfish was stimulated during summer in brackish water, and egg survival and size improved. As a possible explanation of these results, there is a high level of calcium in BRS underground brackish water that causes calcitonin secretion, which positively affects reproduction (Evans, Piermarini, & Choe 2005).

In the BRS/SMC study, sperm concentration in male rainbow trout broodfish held in brackish water for 6 months was significantly higher than in males held in freshwater ($p < 0.01$). Similar results were reported by Atse, Audet, and Noue (2002) for Arctic char, sperm motility did not show any significant difference between broodfish reared in fresh and brackish water ($p > 0.01$), but was slightly higher in brackish water, as confirmed by Landergren and Vallin (1998) working on sea trout, *Salmo trutta*.

TABLE 3 Growth Indices in Different Groups and Salinity Levels

Growth indices/salinity levels	Treatments							
	S				F			
	T1	T2	T3	T4	T5	T6	T7	T8
Survival	99^a	98.1^a	99.5^a	96.2^a	99^a	99.5^a	99^a	96.7^a
WG^1	3.21 ± 0.11^a	3.19 ± 0.41^a	$2.46 + 0.25^b$	2.25 ± 0.17^b	3.69 ± 0.25^a	3.24 ± 0.11^b	3.16 ± 0.01^b	2.65 ± 0.08^c
FCR	0.78 ± 0.01^a	0.84 ± 0.10^a	0.89 ± 0.07^a	0.94 ± 0.07^a	0.77 ± 0.05^a	0.84 ± 0.01^a	0.88 ± 0.03^a	0.91 ± 0.09^a
SGR	6.62 ± 0.08^a	6.59 ± 0.30^a	6.00 ± 0.23^{ab}	$5.80 + 0.17^b$	$6.34 + 0.15^a$	6.05 ± 0.07^b	$5.99 + 0.01^b$	5.60 ± 0.07^c
BWT^2	1340.2 ± 49.7^a	1330.5 ± 172.5^a	1026.3 ± 104.7^{ab}	940.2 ± 71.8^b	1191.3 ± 80.9^a	1046.2 ± 35.6^b	1020.4 ± 4.9^b	856.9 ± 28.9^c

S = the fry originated from broodfish kept in underground brackish water, and F = the fry originated from broodfish kept in freshwater.
T1 = 1 ppt; T2 = 4 ppt; T3 = 7 ppt; T4 = 10 ppt; T5 = 1 ppt; T6 = 4 ppt; T7 = 7 ppt; and T8 = 10 ppt.
Numbers with same letters don't have any difference significantly (p > 0.01).
^{1} = gram; ^{2} = percentage.

The results of our larviculture trial showed that growth indices decreased consistently with increasing salinity. This is in contrast to the findings of Altinok and Grizzle (2003), who reported better growth at intermediate salinity levels (8 to 20 g/l). Morgan and Iwama (1991) supposed that fish might save osmoregulatory energy in isotonic media and thus grow better; this was born out in typical freshwater rainbow trout, but not in steelhead (a sea-run rainbow trout), which did better in pure freshwater. Although Mac Leod (1977) and ourselves found no differences in FCR among larvae reared at different salinities, Altinok and Grizzle (2001) reported that FCR and growth were best at 9 ppt.

Survival in both groups (S&F) was not affected by salinity levels of 1–10 ppt, similar to the findings of Landergren (2001) working with *Salmo trutta* in fresh and slightly brackish water (6 to 7 ppt).

It may be that some of the differences among findings in these studies have to do with the concentration of other elements besides salinity, like calcium, which is higher in concentration in underground waters compared to, and potassium, which has a lower concentration fresh water (Partridge & Lymbery 2008) in underground waters; since, fish extract their potassium requirement directly from water and the potassium requirement relates to the chloride ion concentration, the amount of available potassium should be raised by an increase of salinity.

In conclusion, rainbow trout broodstock rearing in underground brackish water will positively and significantly affect gonad quality and produce bigger larvae at first feeding with higher survival. Although survival was high and better growth of fry from broodstock held in brackish water was observed, fry did better in freshwater overall and more research with stocks of higher initial body weight should be conducted to determine if fish in later life stages might do better in brackish waters (Boeuf & Payan 2001; Morgan & Iwama 1998).

REFERENCES

Albrektsen, S., & O.J. Torrissen. 1988. Physiological changes in blood and seminal plasma during the spawning period of maturing rainbow trout held under different temperature and salinity regimes, and the effect on survival of the broodstock and the eyed eggs. *Inter. Council. Exp. Sea*, 1–24.

Altinok, I., & J.M. Grizzle. 2001. Effects of brackish water on growth, feed conversion and energy absorption efficiency by juvenile euryhaline and freshwater stenohaline fishes. *J. Fish Biol.* 59: 1,142–1,152.

Altinok, I., & J.M. Grizzle. 2003. Effects of low salinities on oxygen consumption of selected euryhaline and stenohaline freshwater fish. *J. World Aquacult. Soc.* 34(1): 113–117.

Atse, C.B., C. Audet, & J.D.L. Noue. 2002. Effects of temperature and salinity on the reproductive success of Arctic char, *Salvelinus alpinus* (L.): Egg composition, milt characteristics, and fry survival. *Aquacult. Res.* 33:299–309.

Atse, C.B. 1999. Evaluation of the impacts of sea level raising on growth, the metabolism and spawning success in *Salvelinus alpinus*. (*Salvelinus alpinus* L.). PhD thesis, Quebec Univeristy at Rimouski, Rimouski, Quebec, Canada.

Boeuf, G., & P. Payan. 2001. Review how should salinity influence fish growth? *Comp. Biochem. Physiol. Part C*, 130:411–423.

Ciereszko, A., & K. Dabrowski. 1995. Sperm quality and ascorbic acid concentration in rainbow trout semen are affected by dietary vitamin C: An across-season study. *Biol. Reprod.* 52:982–988.

Clesceri, L.S., A.E. Greenberg, & A.D. Eaton. 1998. *Standard methods for the examination of water and wastewater*, 20th ed. Washington, DC: American Public Health Association.

Evans, D.H., P.M. Piermarini, and K.P. Choe. 2005. The multifunctional fish gill: dominant site of gas exchange, osmoregulation, acid-base regulation, and excretion of nitrogenous waste. *Physiol Rev.* 85:97–177.

Falahati, M.A. 2003. Comparison study on gonadal development of rainbow trout (*Oncorhynchus mykiss*) in fresh and brackish water. Master's thesis, Tarbiat Modarres University, Tehran, Iran.

Fast, A.W., S.A. Katas, E.G. Grau, & D.K. Barclay. 1991. Seawater maturation and spawning of Rainbow trout in Hawaii. *Progr. Fish-Cult.*, 53:47–49.

Fielder, D.S., W.J. Bardsley, & G.L. Allan. 2001. Survival and growth of Australian snapper, *Pagrus auratus*, in saline groundwater from inland New South Wales, Australia. *Aquacult.* 201:73–90.

Handeland, S.O., B.T. Bjornssonb, A.M. Arnesenc, and S.O. Stefansson. 2003. Seawater adaptation and growth of post-smolt Atlantic salmon (*Salmo salar*) of wild and farmed strains. *Aquacult.* 220:367–384.

Ingram, B.A., L.J. McKinnon, and G.J. Gooley. 2002. Growth and survival of selected aquatic animals in two saline groundwater evaporation basins: an Australian case study. *Aquacult. Res.* 33:425–436.

Landergren, P. 2001. Survival and growth of sea trout parr in fresh and brackish water. *J. Fish Biol.* 58:591–593.

Landergren, P., and L. Vallin. 1998. Spawning of sea trout, *Salmo trutta* L. in brackish water—lost effort or successful strategy? *Fish. Res.* 35:229–236.

Mac Leod, M.G. 1977. Effects of salinity on food intake, absorption and conversion in the rainbow trout *Salmo gairdneri*. *Mar. Biol.* 43:93–102.

Mashaii, N. 2006. *Limnological survey about brackish water pond culture of rainbow trout (Oncorhynchus mykiss). Final report of project*. Tehran, Iran: Iranian Fisheries Research Organization (IFRO).

Morgan, J.D., and G.K. Iwama. 1998. Salinity effects on oxygen consumption, gill Na^+, K^+-ATPase, and ion regulation in juvenile coho salmon. *J. Fish Biol.* 53:1,110–1,119.

Morgan, J.D., & G.K. Iwama. 1991. Effect of salinity on growth, metabolism and ion regulation in juvenile rainbow trout and steelhead trout, *Oncorhynchus mykiss* and fall Chinook salmon, *Oncorhynchus tshawytscha*. *Can. J. Fish. Aquat. Sci.* 48:2,024–2,083.

Nafisi, B.M. 2002. *Culture of rainbow trout (Oncorhynchus mykiss) in earth pond and brackish water of Yazd Province. Final report of project*. Tehran, Iran: Iranian Fisheries Research Organization (IFRO).

Nafisi, B.M., and M.A. Falahati. 2008. *Principle of rainbow trout propagation*. Bushehr, Iran: Persian Gulf University.

Partridge, G.J., & A.J. Lymbery. 2008. The effect of salinity on the requirement for potassium by Barramundi (*Lates calcarifer*) in saline groundwater. *Aquacult.* 278:164–170.

Poole, W.R., and M.G. Dillane. 1998. Estimation of sperm concentration of wild and reconditioned brown trout, *Salmi trutta L. Aquacult. Res.* 29:439–445.

Primary Industries and Resources South Australia (PIRSA). 1999. *Potential for inland saline aquaculture*. Adelaide, South Australia: PIRSA.

The Use of American Ginseng (*Panax quinquefolium*) in Practical Diets for Nile Tilapia (*Oreochromis niloticus*): Growth Performance and Challenge with *Aeromonas hydrophila*

MOHSEN ABDEL-TAWWAB

Department of Fish Biology and Ecology, Central Laboratory for Aquaculture Research, Sharqia, Egypt

This study was carried out to evaluate the effect of American ginseng (AG), Panax quinquefolium, *on growth and resistance to* Aeromonas hydrophila *in Nile tilapia,* Oreochromis niloticus. *Ginseng was included in practical test diets at rates of 0.0 (control), 0.50, 1.0, 2.0, or 5.0 g/kg diet. Fish (9.1 ± 0.3 g) were distributed into quadricated 100-L aquaria at a density of 20 fish per aquarium. Fish in all treatments were fed up to satiation twice daily for 8 weeks. After the feeding trial, fish of each treatment were intraperitoneally injected with pathogenic* A. hydrophila *and kept under observation for 10 days. Highest growth was obtained at 1.0 – 5.0 g AG/kg diet. The survival of fish challenged by* A. hydrophila *increased with increasing AG levels in fish diets. Cost-benefit analysis indicated that ginseng supplementation could reduce per kg costs by 15% with an optimum inclusion level of 2.0 g/kg.*

The author would like to thank Mohamed N. Monier, Department of Fish Biology and Ecology, Central Laboratory for Aquaculture Research (CLAR), Abbassa, Abo-Hammad, Sharqia, Egypt, for his help during the running of this study, and Prof. Azza M. Abdel-Rahman, Department of Fish Disease, CLAR, Abbassa, Abo-Hammad, Sharqia, Egypt, for her help in the bacterial and NBT assays. The author also would like to thank Dr. Randall E. Brummett, Agriculture & Rural Development Department, World Bank, Washington DC, USA, for editing this article.

INTRODUCTION

The application of antibiotics in fish culture is often expensive and undesirable since they could lead to antibiotic resistance and consumer reluctance (Alderman & Hastings 1998; Teuber 2001; Hermann et al. 2003). Moreover, oral chemotherapy may kill or inhibit beneficial microbial flora in the digestive tract (Sugita et al. 1991). Alternatives to the use of antibiotics have been proposed, including medicinal herbs (Xiang & Zhou 2000; Lin et al. 2006; Goda 2008; Won et al. 2008; Abdel-Tawwab et al. 2010; Citarasu 2010). These medicinal herbs could enhance the growth rate of farmed species and anti-microbial activity. Anti-stress characteristics of herbs could enhance the immune system (Citarasu 2010; Harikrishnan et al. 2011). American ginseng (AG), *Panax quinquefolium*, has anti-oxidant activities based on pharmaceutically active saponins, polyacetylenes, polyphenolic compounds, and acidic polysaccharides (Kim et al. 1987; Radad et al. 2006; Luo & Fang 2008). These compounds have been demonstrated to have many functions, including immune modulation (Choi et al. 2008.), inhibition of tumors (Shin et al. 2004), inhibition of pathogenic bacteria (Lee et al. 2006), and anti-peroxidatic reactions (Luo & Fang 2008). The present study was carried out to evaluate the effect of AG on the growth performance of Nile tilapia and its challenge with *Aeromonas hydrophila* infection.

MATERIALS AND METHODS

Five experimental diets were formulated with different levels of *P. quinquefolium*. Ginseng root powder (H.W. Traditional Medicine PTE LTD, 6 Ang Mo Kio Industrial Park 2, Singapore) was included in formulated diets at a rate of 0.0 (control), 0.5, 1.0, 2.0, or 5.0 g/kg diet (Table 1). Ginseng for each diet was suspended in 100 mL water per 1 kg and blended with the other ingredients for 40 min to make a paste of each diet. The pastes were separately passed through a grinder, and pelleted through 1-mm diameter paste extruder. The diets were oven-dried at 55°C for 24 h and stored in plastic bags at −2°C for further use.

Nile tilapia, *O. niloticus*, fingerlings were obtained from nursery ponds at the Central Laboratory for Aquaculture Research, Abbassa, Abo-Hammad, Sharqia, Egypt. Fish were kept in an indoor fiberglass tank for two weeks for adaptation to laboratory conditions. Twenty fish were frozen at −20°C for initial chemical analysis. Fish (9.1 ± 0.3 g) were randomly distributed at a rate of 20 fish per aquarium into twenty 100-L aquaria (four aquaria per treatment).

TABLE 1 Nutrient content and proximate chemical composition (on dry matter basis) of the tested diets differed in ginseng levels.

Ingredients (%)	Ginseng levels (g/kg diet)				
	0.0 (Control)	0.5	1.0	2.0	5.0
Herring fish meal[1]	85	85	85	85	85
Soybean flour[2]	465	465	465	465	465
Wheat bran	183	183	183	183	183
Ground corn	100	100	100	100	100
Corn oil	20	20	20	20	20
Cod liver oil	20	20	20	20	20
Vitamin premix[3]	20	20	20	20	20
Mineral premix[4]	20	20	20	20	20
Starch	87	86.5	86	85	82
Ginseng	0.0	0.5	1.0	2.0	5.0
Total	1000	1000	1000	1000	1000
Chemical composition (%)					
Dry matter	91.2	91.2	91.3	91.2	91.0
Crude protein	30.2	30.2	30.0	30.1	30.5
Ether extract	8.2	8.6	8.7	8.6	8.7
Crude fiber	4.8	4.8	4.5	4.9	4.9
Total ash	6.0	6.0	5.9	6.2	6.1
Nitrogen-free extract[5]	50.8	50.4	50.9	50.2	49.8
Gross energy (kcal/kg)[6]	4569.1	4590.4	4609.2	4576.6	4592.3
P:E ratio	66.10	65.52	65.55	65.77	66.42

[1] Danish fish meal 71.3% protein, 14.2% crude fat, and 11.0% ash obtained from TripleNine Fish Protein, DK-6700 Esbjerg, Denmark.
[2] Egyptian soybean flour 45.6% protein, 6.3% crude fat, and 7.9% obtained from National Oil Co., Giza, Egypt.
[3] Vitamin premix (per kg of premix): thiamine, 2.5 g; riboflavin, 2.5 g; pyridoxine, 2.0 g; inositol, 100.0 g; biotin, 0.3 g; pantothenic acid, 100.0 g; folic acid, 0.75 g; para-aminobenzoic acid, 2.5 g; choline, 200.0 g; nicotinic acid, 10.0 g; cyanocobalamine, 0.005 g; α-tocopherol acetate, 20.1 g; menadione, 2.0 g; retinol palmitate, 100,000 IU; cholecalciferol, 500,000 IU (Juauncey & Ross 1982).
[4] Mineral premix (per kg of premix): $CaHPO_4.2H_2O$, 727.2 g; $MgCO_3.7H_2O$, 127.5 g; KCl 50.0 g; NaCl, 60.0 g; $FeC_6H_5O_7.3H_2O$, 25.0 g; $ZnCO_3$, 5.5 g; $MnCl_2.4H_2O$, 2.5 g; $CuCl_2$, 0.785 g; $CoCl_3.6H_2O$, 0.477 g; $CaIO_3.6H_2O$, 0.295 g; $CrCl_3.6H_2O$, 0.128 g; $AlCl_3.6H_2O$, 0.54 g; Na_2SeO_3, 0.3 g (Juauncey & Ross 1982).
[5] Nitrogen-free extract (NFE) = 100 − (protein % + lipid % + total ash % + crude fiber %).
[6] Gross energy was calculated according to NRC (1993) as 5.65, 9.45, and 4.11 kcal/g for protein, lipid, and carbohydrates, respectively.

Each aquarium was supplied with compressed air. Fish were fed one of the test diets up to satiation twice a day for 8 weeks. Fish in each aquarium were collected, counted, and group-weighed at 2-week intervals. Diets were not offered on sampling days. Settled fish waste along with three-quarters of aquarium's water was siphoned daily and replaced by clean aerated water from a storage tank. Fish mortality was recorded daily and dead fish were removed.

Proximate chemical analyses of test diets (Table 1) and fish samples were done according to standard methods of the Association of Official

Analytical Chemists (1990). Dissolved oxygen and temperature were measured biweekly with an oxygen meter (YSI model 58, Yellow Spring Instrument Co., Yellow Springs, OH, USA). Unionized ammonia (NH_3) was measured using a HANNA kit (HANNA Instruments, Smithfield, RI, USA). The pH was measured by using a pH-meter (Digital Mini-pH Meter, model 55, Fisher Scientific, Denver, CO, USA). In all treatments, water temperature ranged from 27° to 29°C; dissolved oxygen concentrations ranged from 5.4 to 5.7 mg/L; pH ranged from 7.6 to 7.8; and unionized ammonia concentrations ranged from 0.41 to 1.27 mg/L, all of which are within the acceptable range for tilapia growth (Boyd 1984).

Growth performance and feed utilization were calculated as follows:

$$\text{Weight gain (g)} = \text{final weight (g)} - \text{initial weight (g)}$$

$$\text{Specific growth rate (SGR)} = \frac{100\,[\text{Ln final weight (g)} - \text{Ln initial weight (g)}]}{\text{Experimental period (days)}};$$

$$\text{Feed conversion ratio (FCR)} = \text{feed intake (g)/weight gain (g)}$$

After the feeding trial, fish from each treatment were randomly divided into two subgroups; each subgroup contained 10 fish that were stocked into a 100-L aquarium. The challenge test was carried out using *A. hydrophylla* isolated previously at the Department of Fish Disease, Central Laboratory for Aquaculture Research, Abbassa, Abo-Hammad, Sharqia, Egypt. The first subgroup was challenged with pathogenic *A. hydrophila* using a sub-lethal dose as described by Schaperclaus et al. (1992), where a 0.1 ml dose of 24-h broth from virulent *A. hydrophila* (5×10^5 CFU/mL) was injected interperitoneally (IP). The second subgroup was IP injected with 0.1 ml of saline solution as a control. All subgroups were kept under observation for 10 days to record any abnormal clinical signs and the daily fish mortality. Two replicates were carried out for each treatment.

The production of oxygen radicals by leukocytes was assayed by the reduction of nitro blue tetrazolium (NBT, Sigma-Aldrich Chemical, St. Louis, MO, USA) according to Rook et al. (1985). Absorbance was converted to NBT units based on a standard curve of NBT diformazan per milliliter of blood.

The cost of feed to raise a unit of fish biomass was estimated by a simple economic analysis. The estimation was based on local retail market price of all the dietary ingredients at the time of the study. These prices (in Egyptian pound; LE/kg) are as follows: herring fish meal, 12.0; soybean meal, 2.0; wheat bran, 1.4; corn meal, 1.4; corn oil, 9.0; fish oil, 12.0; vitamin premix, 7.0; mineral mixture, 3.0; starch, 2.0; AG 700; and an additional 50.0 LE/ton manufacturing cost (1US$ = 6.1 LE).

Data were subjected to one-way analysis of variance (ANOVA). Differences between means were tested at the 5% probability level using Duncan's new multiple range test. All statistical analyses were done using SPSS V 10 (SPSS, Richmond, VA, USA) as described by Dytham (1999).

RESULTS

Final fish weight, weight gain, and specific growth rate in the AG-fed groups (0.5–5.0 g/kg) were higher than those of the control group ($P < 0.05$; Table 2). The relationship between dietary AG levels and the final weight was best expressed by the second-order polynomial regression equation as follows: $Y = 22.88 - 0.4857X^2 + 3.9143X$ (Figure 1). Fish fed AG-enhanced diets with >1.0 g AG/kg consumed more feed than the other treatments and had a better FCR (1.16, 1.13, and 1.15, respectively; Table 2). Fish survival ranged from 97.8% to 100.0% with no significant difference ($P > 0.05$) among the different treatments. The AG level that produced the most cost-effective fish growth was 2.0 g/kg (Figure 1).

Ginseng supplementation significantly ($P > 0.05$) affected the whole-fish body composition except moisture content, with AG-fed fish featuring slightly higher protein and ash, and slightly lower lipids (Table 3).

Ginseng supplementation stimulated superoxide anion production (NBT values). Highest NBT values were obtained at 2.0–5.0 g AG /kg diet; the lowest value was obtained in the control group. The relationship between dietary AG levels and the NBT values was best expressed by the following equation: $Y = 0.0403 - 0.0132 X^2 + 0.12 X$ (Figure 2).

Fish mortality after injection with *A. hydrophila* increased with time up to a maximum at 4 days post-bacterium injection. Total fish mortality through 10 days after injection declined significantly ($P < 0.05$) with increasing AG level. The lowest fish mortality was obtained at 2.0–5.0 g AG/kg diet. The relationship between dietary AG levels and the cumulative fish mortality after bacterial challenge for 10 days was best expressed by the following equation: $Y = 95.2 + X^2 - 11.6 X$ (Figure 3).

The AG-containing diets were economically superior to the control diet up to 2.0 g/kg diet, which reduced the cost to produce ton fish gain by 15.2% (Table 4).

DISCUSSION

Similar results to those observed here were obtained by Goda (2008), who reported that ginseng, *Panax ginseng*, significantly improved Nile tilapia growth and feed utilization. The changes in fish body composition due to AG supplementation suggest that the dietary AG plays a role in enhancing

TABLE 2 Growth performance and feed utilization of Nile tilapia fed a 30% CP diet containing different levels of ginseng for 8 weeks.

Items	Ginseng levels (g/kg diet)				
	Control (0.0)	0.5	1.0	2.0	5.0
Initial weight (g)	9.1 ± 0.03	9.1 ± 0.03	9.2 ± 0.07	9.2 ± 0.09	9.1 ± 0.09
Final weight (g)	26.0 ± 0.52 c	28.3 ± 0.41 b	29.8 ± 0.90 a	30.4 ± 0.84 a	30.7 ± 1.01 a
Weight gain (g)	16.9 ± 0.50 c	19.2 ± 0.38 b	20.6 ± 0.96 a	21.2 ± 0.78 a	21.6 ± 0.94 a
SGR (%/day)	1.88 ± 0.032 c	2.03 ± 0.021 b	2.10 ± 0.062 a	2.13 ± 0.042 a	2.17 ± 0.049 a
Feed intake (g feed/fish)	22.1 ± 0.23 c	23.6 ± 0.33 b	23.9 ± 0.12 ab	24.0 ± 0.19 ab	24.8 ± 0.28 a
Food conversion ratio	1.31 ± 0.08 a	1.23 ± 0.08 ab	1.16 ± 0.06 b	1.13 ± 0.04 b	1.15 ± 0.08 b
Survival rate (%)	100 ± 0.00	100 ± 0.00	97.8 ± 2.23	97.8 ± 2.23	100 ± 0.00

Means having the same letter in the same row are not significantly different at $P < 0.05$.

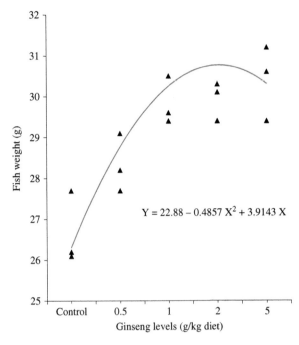

FIGURE 1 The final weight (g) of Nile tilapia fed diets containing different ginseng levels for 8 weeks. (Color figure available online.)

TABLE 3 Proximate chemical analysis (%; on dry matter basis) of whole-body Nile tilapia fed a 30% CP diet containing different levels of ginseng for 8 weeks.

	Ginseng levels (g/kg diet)				
Items	Control (0.0)	0.5	1.0	2.0	5.0
Moisture	72.5 ± 0.31	72.7 ± 0.70	73.4 ± 0.57	73.1 ± 0.39	73.7 ± 0.59
Crude protein	63.9 ± 1.02 b	63.7 ± 1.08 b	64.6 ± 1.08 b	66.7 ± 0.50 ab	67.3 ± 0.71 a
Ether extract	17.8 ± 1.64 a	15.4 ± 1.42 ab	15.6 ± 1.65 ab	13.4 ± 1.16 b	12.2 ± 0.57 b
Total ash	16.3 ± 0.77 b	17.8 ± 0.79 b	18.3 ± 1.60 ab	18.8 ± 0.56 ab	19.2 ± 0.58 a

Means having the same letter in the same row are not significantly different at $P < 0.05$.

feed intake with a subsequent enhancement of nutrient deposition in fish body. On the other hand, changes in protein and lipid content in the fish body could be linked with changes in their synthesis, deposition rate in muscle, and/or different growth rate (Fauconneau 1985; Abdel-Tawwab et al. 2006). In his study, Goda (2008) found no significant changes in whole-body proximate analysis of Nile tilapia due to ginseng supplementation.

The observation that enhanced resistance to *A. hydrophila* infection and increased NBT values associated with increasing AG levels suggest that AG acts as an immuno-stimulant. This may be attributed to one or more of its constitutes, saponins, or ginsenosides, the principle active ingredients in

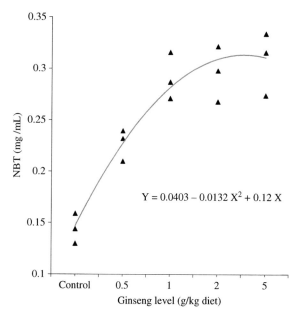

FIGURE 2 Changes in nitro blue tetrazolium (NBT; mg/mL) in Nile tilapia fed diets containing different ginseng levels for 8 weeks. (Color figure available online.)

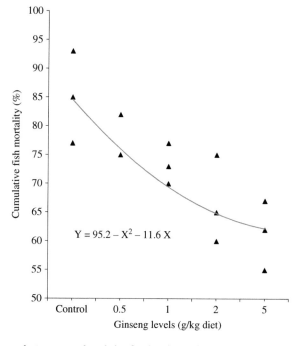

FIGURE 3 The cumulative mortality (%) of Nile tilapia fed diets containing different ginseng levels for 8 weeks and challenged by *Aeromonas hydrophila* for 10 days. (Color figure available online.)

TABLE 4 Economic efficiency for production of 1 kg gain of Nile tilapia fed a 30% CP diet containing different levels of ginseng for 8 weeks.

	Ginseng levels (g/kg diet)				
	Control (0.0)	0.5	1.0	2.0	5.0
Feed cost (L.E./kg)	3.14	3.14	3.15	3.15	3.17
FCR (kg feed/kg gain)	1.31	1.23	1.16	1.13	1.15
Feed cost per ton gain (L.E.)	4,179.2	3,927.0	3,706.4	3,616.2	3,697.5
Cost reduction per ton gain (L.E.)*	0.0	252.1	472.7	562.9	481.7
Cost reduction per ton gain (%)**	0.0	6.0	12.0	15.2	13.3

*Cost reduction per kg gain (L.E.) = feed cost per ton gain of control (L.E.) − feed cost per ton gain of AG-containing diets (L.E.).
**Cost reduction per kg gain (%) = 100 (cost reduction per ton gain [L.E.] in AG-containing diets/feed cost per ton gain of control [L.E.]).

ginseng (Liu & Xiao 1992; Back et al. 1996). Ginsenosides are unique to *Panax* species, many of which exist in minute amounts and are believed to be responsible for most of ginseng's anti-oxidant activity (Tyler 1993; Attele et al. 1999; Wakabayashi et al. 1997). Furthermore, Liu et al. (2011) found that ginseng polysaccharides have anti-oxidant activity in whiteleg shrimp, *Litopenaeus vannamei*. The usefulness of antioxidants in protecting cellular components against oxidative stress is well established (Mohan et al. 2006).

Similar results were found by Won et al. (2008) who studied the effects of Siberian ginseng, *Eleutherococcus senticosus*, residuum extract (SG-RE) on non-specific immunity in olive flounder, *Paralichthys olivaceus*, where fish were fed diets supplemented with SG-RE (0%, 3%, or 7%) for 8 weeks. They found that fish fed a diet supplemented with 3% SG-RE exhibited improved innate immunity and developed resistance to *Edwardsiella tarda* and *Vibrio anguillarum* infection.

REFERENCES

Abdel-Tawwab, M., M. H. Ahmad, M. E. A. Seden, and S. M. F. Sakr. 2010. Use of green tea, *Camellia sinensis* L., in practical diets for growth and protection of Nile tilapia, *Oreochromis niloticus* (L.), against *Aeromonas hydrophila* infection. *Journal of the World Aquaculture Society* 41:203–213.

Abdel-Tawwab M., Y. A. E. Khattab, M. H. Ahmad, and A. M. E. Shalaby. 2006. Compensatory growth, feed utilization, whole body composition, and hematological changes in starved juvenile Nile tilapia, *Oreochromis niloticus* (L.). *Journal of Applied Aquaculture* 18:17–36.

Alderman, D. J., and T.S. Hastings. 1998. Antibiotic use in aquaculture: development of antibiotic resistance-potential for consumer health risk. *International Journal of Food Science and Technology* 33:139–155.

Association of Official Analytical Chemists (AOAC). 1990. *Official methods of analysis of the Association of Official Analytical Chemists*, 15th ed. Arlington, VA: AOAC.

Attele, A. S., J. A. Wu, and C. S. Yuan. 1999. Ginseng pharmacology: multiple constituents and multiple actions. *Biochemical Pharmacology* 58:1,685–1,693.

Back, N. I., D. S. Kim, Y. H. Lee, J. D. Park, C. B. Lee, and S. I. Kim. 1996. Ginsenoside Rh4, a genuine dammarane glycoside from Korean red ginseng. *Planta Medica* 62:86–87.

Boyd, C. E. 1984. *Water quality in warmwater fishponds*. Auburn, AL: Auburn University Agricultural Experimental Station.

Choi, H., K. Kim, and E. Sohn. 2008. Red ginseng acidic polysaccharide (RGAP) in combination with IFN-r results in enhanced macrophage function through activation of the NF-kB pathway. *Bioscience, Biotechnology, and Biochemistry* 72:1,817–1,825.

Citarasu, T. 2010. Herbal biomedicines: A new opportunity for aquaculture industry. *Aquaculture International* 18:403–414.

Dytham C. 1999. *Choosing and using statistics: A biologist's guide*. London: Blackwell Science.

Fauconneau, B. 1985. Protein synthesis and protein deposition in fish. In *Nutrition and feeding in fish*, edited by C. B. Cowey, A. M. Mackie, and J. G. Bell, 17–45. London: Academic Press.

Goda, A. M. A.-S. 2008. Effect of dietary ginseng herb (Ginsana_G115) supplementation on growth, feed utilization, and hematological indices of Nile tilapia, *Oreochromis niloticus* (L.), fingerlings. *Journal of the World Aquaculture Society* 39:205–214.

Harikrishnan, R., C. Balasundaram, and M.-S. Heo. 2011. Impact of plant products on innate and adaptive immune system of cultured finfish and shellfish. *Aquaculture* 317:1–15.

Hermann, J. R., M. S. Honeyman, J. J. Zimmerman, B. J. Thacker, P. J. Holden, and C. C. Chang. 2003. Effect of dietary *Echinacea purpurea* on viremia and performance in porcine reproductive and respiratory syndrome virus-infected nursery pigs. *Journal of Animal Science* 81:2,139–2,144.

Kim, M. W., S. R. Ko, K. J. Choi, and S. C. Kim. 1987. Distribution of saponin in various sections of *Panax ginseng* root and changes of its contents according to root age. *Korean Journal of Ginseng Science* 11:10–16.

Lee, J., J. S. Shim, and J. S. Lee. 2006. Pectin-like acidic polysaccharide from *Panax ginseng* with selective anti-adhesive activity against pathogenic bacteria. *Carbohydrate Research* 341:1,154–1,163.

Lin, H.-Z., Z.-J. Li, Y.-Q. Chen, W.-H. Zheng, and K. Yang. 2006. Effect of dietary traditional Chinese medicines on apparent digestibility coefficients of nutrients for white shrimp, *Litopenaeus vannamei*, Boone. *Aquaculture* 253:495–501.

Liu, C. X., and P. G. Xiao. 1992. Recent advances on ginseng research in China. *Journal of Ethnopharmacology* 36:27–38.

Liu, X. L., Q. Y. Xi, L. Yang, H. Y. Li, Q. Y. Jiang, G. Shu, S. B. Wang, P. Gao, X. T. Zhu, and Y. L. Zhang. 2011. The effect of dietary *Panax ginseng* polysaccharide extract on the immune responses in white shrimp, *Litopenaeus vannamei*. *Fish and Shellfish Immunology* 30:495–500.

Luo, D. H., and B. S. Fang. 2008. Structural identification of ginseng polysaccharides and testing of their antioxidant activities. *Carbohydrate Polymers* 72:376–381.

Mohan, I. K., M. Khan, J. C. Shobha, M. U. Naidu, A. Prayag, and P. Kuppusamy. 2006. Protection against cisplatin-induced nephrotoxicity by *Spirulina* in rats. *Cancer Chemistry and Pharmacology* 58:802–808.

National Research Council (NRC). 1993. *Nutrient requirements of fish*. Washington DC: National Academy Press.

Radad, K., G. Gille, L. Linlin, and W. D. Rausch. 2006. Use of ginseng in medicine with emphasis on neurodegenerative disorders. *Journal of Pharmacological Science* 100:175–186.

Rook, G. A. W., J. Steele, S. Umar, and H. M. Dockrell. 1985. A simple method for the solubilisation of reduced NBT, and its use as a colorimetric assay for activation of human macrophages by γ-interferon. *Journal of Immunological Methods* 82:161–167.

Schaperclaus, W., H. Kulow, and K. Schreckenbach. 1992. *Fish disease*. Rotterdam, The Netherlands: A.A. Balkema.

Shin, H., Y.S. Kim, and Y. Kwak. 2004. A further study on the inhibition of tumor growth and metastasis by red ginseng acidic polysaccharide (RGAP). *Natural Product Sciences* 10:284–288.

Sugita, H., C. Miyajima, and Y. Deguchi. 1991. The vitamin B12-producing ability of the intestinal microflora of freshwater fish. *Aquaculture* 92:267–276.

Teuber, M. 2001. Veterinary use and antibiotic resistance. *Current Opinion in Microbiology* 4:493–499.

Tyler, V. E. 1993. *The honest herbal: A sensible guide to the use of herbs and related remedies*, 3rd ed. New York: Haworth Press.

Wakabayashi, C., H. Hasegawa, J. Murata, and I. Saiki. 1997. In vivo anti-metastatic action of ginseng protopanaxadiol saponins is based on their intestinal bacterial metabolism after oral administration. *Oncology Research* 9:411–417.

Won, K. M., P. K. Kim, S. H. Lee, and S. I. Park. 2008. Effect of the residuum extract of Siberian ginseng, *Eleutherococcus senticosus* on non-specific immunity in olive flounder, *Paralichthys olivaceus*. *Fisheries Science* 74:635–641.

Xiang, X., and X. H. Zhou. 2000. Application effect of Chinese herb medicine to aquatic animal feeds. *Cereal Feed Index* 3:27–29.

Use of Spray-Dried Blood Meal as an Alternative Protein Source in Pirarucu (*Arapaima gigas*) Diets

RICARDO AMARAL RIBEIRO[1], RODRIGO OTÁVIO DE ALMEIDA OZÓRIO[2], SÓNIA MARIA GOMES BATISTA[2], MANOEL PEREIRA-FILHO[3], EDUARDO AKIFUMI ONO[3], and RODRIGO ROUBACH[3]

[1] *Universidade Federal do Acre, Departamento de Ciências Agrárias, Rio Branco, Brazil*
[2] *Centro Interdisciplinar de Investigação Marinha e Ambiental–Universidade do Porto (CIMAR/CIIMAR), Oporto, Portugal*
[3] *Instituto Nacional de Pesquisas da Amazônia, Coordenação Pesquisas Aquicultura, Manaus, Amazonas, Brazil*

*We evaluated the effects of various dietary blood meal levels on the growth performance and body composition of pirarucu (*Araipama gigas*) juveniles. Fish (8.5 ± 0.4 g) were stocked into 24 tanks and fed for 60 days with eight isoproteic diets, having 0% to 21% incorporation of blood meal. Fish increased weight by six to 15 times from their initial weight. The highest body weights (117–135 g), growth rates (4%–5% BW/day), and protein retentions (19%–20%) were observed in fish fed 0% to 6% blood meal. Feed intake increased from 2.9% to 4.4% BW/day with increasing blood meal level. Feed conversion ratio (FCR: 1.0–1.1) did not vary among groups fed 0% to 6% blood meal ($P < 0.001$). However, when fish were fed more than 6% blood meal, FCR (1.3–1.7) and protein retention (11.1–13.7) deteriorated sharply. Lipid (7.7–11.7%) and energy (18.7–21.2 kJ/g) content increased with increasing blood meal levels up to 9%.*

We are thankful to Ms. Maria Inês Pereira, Mr. Rui Sant'Ana, Ms. Cydia Furtado, and Mr. Rondon for their valuable assistance during the feeding trial and chemical analyses.

INTRODUCTION

The pirarucu (*Arapaima gigas*) is one of the largest species among the South America freshwater fishes (Fontenele 1948; Pontes 1977). A compulsory air-breather of the family Osteoglossidae (Nelson 1994; Li & Wilson 1996), it is commonly found in floodplain systems from Brazil, Peru, Colombia, and Guiana, where they inhabit large lakes and slow-running rivers. Decline in natural populations of pirarucu due to overfishing and habitat change have prompted private and governmental organizations to finance research to produce juveniles for restocking (Saint-Paul 1986; Queiroz & Sardinha 1999; Viana et al. 2004; Castello, Viana, & Watkins 2005) and for meat production (Crossa & Chaves 2005). It is an excellent candidate species for aquaculture due to its high market price, very high growth rates, reproduction in captivity, and tolerance to poor water quality (Cavero et al. 2003a). The piscivorous feeding habit of adult pirarucu is, however, a bottleneck for its production in captivity. Its low acceptance to formulated feed (Kubitza 1995; Hayashi et al. 1999) can be overcome through feed training of juvenile pirarucu, gradually replacing live food by inert feed (Crescencio 2001; Cavero et al. 2003a; Silva 2004; Scorvo-Filho et al. 2004).

Pereira-Filho et al. (2003) obtained biomass growth of 5,298% in 12 months on extruded diets (40% crude protein, 3.400 Kcal/kg diet). Cavero et al. (2003b) testing the effect of stocking density (15, 20, and 25 fish/m^3) on the growth performance of pirarucu (initial body weight, 11 g; final body weight, 108 g) fed on extruded diets (45% crude protein) showed that feed conversion (0.8) and condition factor were improved at higher densities. In addition, aggressive behavior (cannibalism and feed competition) was significantly reduced with increasing stocking density.

However, fishmeal costs for these diets are high, and there is an urgent need to find alternatives. Slaughterhouse wastes, including blood meal, is one potential protein substitute for fishmeal. Currently, the state of the Acre in the northern region of Brazil possesses a cattle herd of 2.1 million animals, in which 250,000 animals are killed yearly. All blood produced in the slaughterhouses is deposited in stabilization lagoons, and eventually most of it is discarded. For an environmental and aquaculture viewpoint, it is desirable that the blood stored in such lagoons can be reused as a feed ingredient by drying it to produce blood meal (Alan et al. 2000). Gomes da Silva and Oliva-Teles (1998) reported that spray-dried blood meal was an effective protein source in fish diet. Previous research has shown that blood meal incorporated up to a level of 6% in the diets of groupers, *Epinephelus coioides* (Millamena 2002), and tambaqui, *Colossoma macropomum* (Martins & Guzman 1994), resulted in similar growth performance when compared with those fed fishmeal-based diets. However, problems with palatability and variation in amino acid availability may limit the use of spray-dried blood meals in fish diets.

The aim of the current study is to evaluate whether blood meal, a less expensive and locally available protein source than fishmeal, can efficiently meet the protein needs of pirarucu.

MATERIALS AND METHODS

A total of 192 pirarucu juveniles from a single family and with identical nutritional history were obtained from a commercial fish hatchery. The feeding trial was carried out in the research facility of the Agrarian Sciences Department of the Federal University of Acre, Brazil.

After two weeks acclimatization, fish averaging 8.5 ± 0.4 g were randomly distributed into 24 circular fiberglass tanks of 250-L each (eight fish/tank), connected to a flow-through system. Fish were fed one of eight extruded diets, identical in digestible energy/crude protein ratio (DE/CP, 9 kcal/g) with different levels of spray-dried, sterilized, powdered bovine blood (Polinutre Alimentos Ltd., São Paulo, Brazil) of 0%, 3%, 6%, 9%, 12%, 15%, 18%, and 21% (Table 1). Fishmeal, meat and bone meal, and poultry fat contents were adjusted to keep the diets with similar digestible energy to protein ratios.

For diet preparation, dry ingredients were homogeneously ground to 500 μm, thoroughly mixed, and water was added to about 50%. Mixtures were cold extruded through a 5.0-mm die mincer and dried for 24 h in a forced-air oven (55°C). Dried diets were crumbled and sieved into 1 to 3 mm pellets, a procedure necessary during the first 30 days of the feeding trial. After that fish were large enough to feed on 5.0-mm diet. All diets were kept frozen (−10°C) until immediately before use.

Proximate composition of the diets and carcass were carried out by the National Institute for Research in the Amazon (INPA), Laboratory of Fish Nutrition. Dry matter (4 h, 105°C) and ash (5 h, 550°C) were determined on fresh matter basis. Crude fat (petroleum ether extraction, Soxhlet method, 40–60°C), crude protein (macro-Kjeldahl; N × 6.25), and gross energy (adiabatic bomb calorimeter, IKA C2000) were determined on freeze-dried material.

Diets were randomly assigned to triplicate tanks and hand-fed *ad libitum*, seven days a week. Uneaten pellets were removed after 20 min for determination of consumption. Feeding frequency was gradually reduced from four to two times per day as the fish grew (Gandra 2002): four times per day (day 0 to day 10: 07:00, 11:00, 15:00, and 19:00 h), three times per day (day 11 to day 20: 08:00, 12:00, and 18:00 h) and two times per day (day 21 to day 60: 08:00 and 17:00 h).

TABLE 1 Feed Formulation and Chemical Composition of the Experimental Diets (Percentage of as-Fed Basis)

	Dietary blood meal level (%)							
	0	3	6	9	12	15	18	21
Ingredients (%)								
Fishmeal[1]	42	39	36	32.5	29	25	21.5	18
Blood Meal[2]	0	3	6	9	12	15	18	21
Soymeal	21	21	21	21	21	21	21	21
Wheat	19	19	20	20	20	20.5	20.5	20.5
Meat and Bone Meal[3]	7	6.5	5.5	5.5	5.5	5.5	5.5	5.5
Poultry Fat	10	10.5	10.5	11	11.5	12	12.5	13
Premix[4]	1	1	1	1	1	1	1	1
Nutrients								
Dry Matter (%)	96.4	95.8	95.8	96.0	94.4	96.3	96.6	94.5
Crude Protein (%)	53.4	53.5	53.6	54.3	57.5	55.5	55.8	56.6
Crude Lipid (%)	15.4	15.5	15.1	14.2	15.4	15.5	15.7	15.8
Ash (%)	11.8	10.9	10.5	9.8	9.6	8.4	8.3	8
DE/CP (kcal/g)	9	9	8.9	9	9	9	9	9
Essential Amino Acids (Calculated)								
Lysine	2.90	2.91	2.89	2.62	2.85	2.81	2.79	2.77
Methionine	1.01	0.98	0.93	0.84	0.84	0.79	0.74	0.70
Arginine	1.97	1.96	1.93	1.90	1.88	1.83	1.81	1.78
Histidine	0.70	0.76	0.82	0.87	0.92	0.97	1.03	1.08
Isoleucine	1.33	1.30	1.26	1.22	1.17	1.12	1.08	1.04
Leucine	2.33	2.39	2.45	2.50	2.55	2.58	2.63	2.68
Valine	1.60	1.64	1.69	1.73	1.77	1.79	1.83	1.87
Phenylalanine	1.23	1.26	1.30	1.33	1.36	1.38	1.41	1.44
Threonine	1.72	1.71	1.71	1.67	1.69	1.67	1.67	1.66
Tryptophan	0.47	0.51	0.54	0.52	0.61	0.64	0.68	0.72

[1] Dry matter, 89.5%; crude protein, 67%; crude lipid, 10%; Lys, 5.0%; Met, 1.9%; Thr, 2.8%; and Trp, 0.67%.
[2] Blood meal spray dried (dry matter, 91.3%; crude protein, 83%; crude lipids, 0.5%; ash, 1.5%; energy, 4.9 MCal/kg; Lys, 5.17%; Met, 0.71%; Phe, 3.82%; Arg, 3.7%; His, 3.45%; Iso 1.76%; Leu, 7.22%; Thr, 3.04%; Val, 5.15%; abd Trp, 1.96%).
[3] Dry matter, 95%; crude protein, 45.6%; crude lipid, 11%; and ash, 33%.
[4] Contained per kg mixture: vitamin: A, 6,000,000 IU; B_1, 5000 mg; B_2, 1,120 mg; B_3, 30,000 mg; B_5, 30,000 mg; B_6, 8,000 mg; B_8 2,000 mg; B_9, 3,000 mg; B_{12} 20,000 mcg; C, 500 mg; D_3, 2,250,000 IU; K_3, 3,000 mg; E, 75,000 mg. Minerals: $ZnSO_4$, 150,000; $MnSO_4$, 60,000; KI, 4,500; $FeSO_4$, 100,000 mg; $CoSO_4$ 2,000 mg; and Na_2SeO_3, 400 mg.

Dissolved oxygen (4.0 ± 0.6 mg/L), pH (6.4 8 ± 0.17), total ammonia nitrogen (0.61 ± 0.4 mg/L), and nitrite (0.07 ± 0.06 mg/L) were monitored weekly. Water temperature (25.2 ± 1.2°C) was checked daily. Photoperiod was a natural 12 L:12 D in the tropics.

Ten fish at the start and two fish/tank (n = 6) at the conclusion of the feeding trial were sacrificed by an anesthetic overdose (Benzocaine; 0.2 g L^{-1} H_2O) and weighed to the nearest 0.1 g. Fish from each tank were pooled together and immediately stored at −20°C. Frozen fish were minced and freeze-dried for subsequent proximate analyses.

Performance and feed efficiency were determined as individual weight gain, feed conversion ratio (feed intake/weight gain, g/g), voluntary fed intake (% body weight/day), daily growth index (%/day), protein efficiency ratio (g/g), protein retention, and condition factor, according to the following equations:

Daily growth index (DGI) = $[(100)(FBW^{1/3} - IBW^{1/3})/\text{trial duration}]$,

where trial duration is 60 days.

Voluntary feed intake (VFI) = $[(100)(\text{crude feed intake}/ABW/\text{trial duration})]$,

where ABW = (IBW + FBW)/2.

Feed conversion ratio (FCR) = weight gain/feed intake.

Protein efficiency ratio (PER) = weight gain/protein intake.

Condition factor (K) = $[(100)(FBW, g)/(Lt^3, cm)]$,

where Lt is the final standard length.

Nutrient retention = $[(100) \times (\text{nutrient gain})/(\text{nutrient intake})]$

One-way analysis of variance (ANOVA) was performed, followed by Tukey's multiple range test. Data were expressed as the mean ± SD of three replicates, using each tank as the experimental unit. Differences were considered significant if P was less than 0.05. All statistical analyses were performed using ASSISTAT (2008).

RESULTS

Growth performances and nutrient utilization were significantly affected by the different dietary blood meal levels. Fish fed above 12% blood meal had reduced weight gain and lower growth rates (DGI) when compared with fish fed 0%, 3%, 6%, and 9% blood meal incorporation (P < 0.0001) (Table 2). Fish fed up to 6% blood meal showed the lowest feed conversion ratio (FCR, 1.0–1.1) when compared to fish fed 9% blood meal or more (1.3–1.7). Feed intake (VFI, % BW/d) and protein intake increased from 2.9% to 4.4%/day and 62–74 g/fish, respectively (P < 0.001), while protein efficiency rate (PER) decreased linearly from 1.86 to 1.0 g/g with increasing dietary blood meal level (P < 0.0001). Condition factor (K) varied between 5.2 and 5.8, but values were not statistically different among treatments.
Protein retention was significantly higher in fish fed up to 6% blood meal (19.3%–20.0% intake) when compared to those fed 9% to 21% blood meal

TABLE 2 Effect of Different Dietary Blood Meal Levels on Weight Gain, Feed Conversion, and Nutrient Utilization in Juvenile Pirarucu Fed Over 60 Days

	Dietary blood meal level (%)							
	0	3	6	9	12	15	18	21
Initial Body Weight (IBW, g)	8.5 ± 0.3	8.3 ± 0.3	8.2 ± 0.6	8.6 ± 0.5	8.3 ± 0.2	8.5 ± 0.5	8.7 ± 0.4	8.6 ± 0.1
Final Body Weight (FBW, g)	135.2 ± 3.5^a	122.7 ± 11.2^{ab}	117.0 ± 11.8^{ab}	115.3 ± 13.7^{ab}	105.5 ± 9.5^{bc}	89.2 ± 1.5^{cd}	74.1 ± 1.8^d	72.4 ± 3.0^d
Daily Growth Index[1] (%BW/day)	5.03 ± 0.31^a	4.52 ± 0.27^{ab}	4.79 ± 0.22^a	4.44 ± 0.46^{ab}	4.38 ± 0.44^{ab}	3.82 ± 0.48^{bc}	3.58 ± 0.07^{bc}	3.31 ± 0.26^c
Voluntary Feed Intake[2] (%BW/day)	2.94 ± 0.13^c	3.14 ± 0.14^c	3.00 ± 0.22^c	3.75 ± 0.09^b	3.75 ± 0.09^b	4.07 ± 0.13^{ab}	4.35 ± 0.26^a	4.36 ± 0.14^a
Feed Conversion Ratio[3] (g/g)	1.01 ± 0.05^c	1.10 ± 0.07^c	1.04 ± 0.08^c	1.33 ± 0.07^b	1.33 ± 0.01^b	1.52 ± 0.12^{ab}	1.65 ± 0.08^a	1.71 ± 0.09^a
Protein Intake (g/fish)	62.36 ± 3.46^{abc}	58.09 ± 1.91^c	60.3 ± 1.95^{bc}	71.82 ± 8.46^{ab}	74.07 ± 2.36^a	63.26 ± 1.27^{abc}	57.53 ± 3.58^c	56.15 ± 2.05^c
Protein Efficiency Ratio[4]	1.86 ± 0.10^a	1.70 ± 0.11^a	1.80 ± 0.14^a	1.38 ± 0.08^b	1.31 ± 0.01^{bc}	1.19 ± 0.09^{bcd}	1.09 ± 0.05^{cd}	1.04 ± 0.06^d
Condition Factor	5.23 ± 0.56	5.14 ± 0.53	5.81 ± 0.47	5.43 ± 0.85	5.48 ± 0.63	4.95 ± 0.67	5.49 ± 0.29	5.23 ± 1.15
Nutrient Retention								
Protein[6] (%)	20.0 ± 2.3^a	18.6 ± 1.9^a	19.3 ± 0.8^a	11.8 ± 1.7^b	13.7 ± 1.2^b	12.9 ± 0.9^b	11.1 ± 0.6^b	11.1 ± 0.8^b
Lipid[6] (%)	6.7 ± 1.5^b	9.0 ± 1.1^{ab}	10.8 ± 0.1^a	8.7 ± 1.7^{ab}	7.9 ± 0.6^{ab}	7.2 ± 1.1^b	7.1 ± 1.3^b	6.1 ± 0.4^b

Values (mean ±SD n=3) sharing common superscripts within rows are not significantly different (P > 0.05).
[1]DGI: [(100)(FBW$^{1/3}$ - IBW$^{1/3}$)/trial duration], where trial duration = 60 days.
[2]VFI = [(100)(crude feed intake/ABW/trial duration)], where ABW = (IBW + FBW)/2.
[3]FCR = feed intake/weight gain.
[4]PER = weight gain/protein intake.
[5]K = [(100)(FBW, g)/(Lt3, cm)], where Lt is the final standard length.
[6]Nutrient retention = [(100)x(nutrient gain)/(nutrient intake)]

(13.7%–11.1% intake). Lipid retention increased linearly from 7% to 11% intake as dietary blood meal level increased from 0% to 6% incorporation, and then leveled off to 6% intake.

The whole body composition of fish before and after the experiment is shown in Table 3. Moisture (83.8%–86.3%) and protein (64%–68%) did not vary among treatments. There were statistical differences in lipid ($P < 0.03$), ash ($P < 0.002$), and energy ($P < 0.034$) content among treatments. Lipid (7.7–11.7%) and energy (18.7–21.2 kJ/g) content increased linearly with increasing blood meal level from 0% to 9% incorporation. Ash decreased from 15.6% to 12.7% as dietary blood meal level increased from 0 % to 18%.

DISCUSSION

Results indicate that growth performance is inhibited when pirarucu are fed diets supplemented with more that 6% blood meal. Similar results were obtained for gilthead sea bream, S*parus aurata* (Martínez-Llorens et al. 2008). Their results demonstrated that blood meal can partially substitute for fishmeal (up to 10%) in sea bream diets with no negative effect on performance. Luzier, Summerfelt, and Ketola (1995) showed that juvenile trout performed well when fed diets with up to 20% blood meal incorporation, while Murray cod, *Maccullochella peelii peelii*, showed retarded growth and high mortalities when fed diets supplemented with blood meal (Abery, Gunasekera, & De-Silva 2002).

The differences among these results may have been related to the quality and processing of the blood meal used, species, fish size, duration of the experiment, and culture systems. Blood meal is produced using a wide variety of processing techniques (Bureau, Harris, & Cho 1999). Many studies indicated that time of exposure to heating, temperature of processing, and drying are critical factors determining the digestibility of protein and amino acids of blood meal and other heat-treated ingredients. Zhou et al. (2008) working with bluntnose black bream, *Megalobrama amblycephala*, observed that dried blood meal had low protein digestibility (23.7%). In comparison, the spray-dried blood meal used in this trial, similar to that used by Martínez-Llorens et al. (2008), may have protein digestibility above 90%.

Some animal by-products can lower the palatability of diets, resulting in reduced intake and poor growth. The reduced weight gain and lower daily growth rates (DGI, %BW/d) observed in fish fed more than 12% was possibly an effect of low palatability of blood meal. Low diet palatability has been demonstrated as responsible for the reduced growth of fish fed diets in which high levels of fishmeal were replaced by economically more accessible and locally available plant or animal protein ingredients (Davis, Jirsa, & Arnold 1995; Xue & Cui 2001).

TABLE 3 Effect of Different Dietary Blood Meal Levels on Whole Body Composition (% Dry Matter Basis) of Juvenile Pirarucu Fed Over 60 Days

	Initial	Dietary blood meal level (%)							
		0	3	6	9	12	15	18	21
Moisture (%)	85.4 ± 0.2	85.3 ± 2.6	83.8 ± 1.4	84.2 ± 0.7	86.3 ± 1.9	84.5 ± 1.0	84.1 ± 0.7	84.4 ± 0.8	84.4 ± 0.2
Crude protein (%)	75.6 ± 5.7	65.5 ± 2.3	67.6 ± 3.0	67.9 ± 1.4	64.0 ± 1.7	68.0 ± 1.9	67.8 ± 2.3	66.4 ± 0.5	68.1 ± 0.8
Crude lipid (%)	2.9 ± 0.3	7.7 ± 2.0^b	9.3 ± 0.6^{ab}	10.5 ± 0.4^{ab}	11.7 ± 2.1^a	10.4 ± 0.1^{ab}	10.1 ± 0.7^{ab}	11.2 ± 2.4^a	9.9 ± 0.1^{ab}
Ash (%)	18.2 ± 2.4	15.6 ± 1.9^a	15.2 ± 0.7^{ab}	14.5 ± 0.5^{abc}	13.1 ± 0.7^{bc}	14.0 ± 0.7^{bc}	13.2 ± 0.8^{bc}	12.7 ± 0.9^c	13.6 ± 0.9^{bc}
Energy (kJ/g)	–	18.7 ± 0.0^b	20.4 ± 0.7^{ab}	19.8 ± 0.2^{ab}	21.2 ± 1.2^a	20.1 ± 0.4^{ab}	19.9 ± 0.0^{ab}	19.6 ± 0.4^{ab}	20.3 ± 0.7^{ab}

In the current study, fish fed 9% blood meal or more showed a higher feed conversion. Guo, Wang, and Bureau (2007), testing several rendered animal ingredients, including blood meal, to replace fishmeal in diets for cuneate drum, *Nibea miichthioides*, observed that a blend of such ingredients provided a nutritionally balanced and cost-effective diet when compared to using them individually. The protein quality of a diet is determined by the total amino acid balance and their bioavailability (Boorman 1992). In fact, methionine intake in the present study was lower in fish fed blood meal diets. This was expected since blood meal is deficient in this amino acid. Probably the use of blood meal in combination with a methionine-rich protein source, such as Brazil nut, *Bertholletia excelsa* (Souza & Menezes 2004), could allow higher inclusion levels of blood meal with no adverse effects on performances.

Moreover, voluntary feed intake (VFI) increased linearly with increasing dietary blood meal level. In addition, the ability to retain dietary nutrients decreased in fish fed 9% to 21% blood meal when compared to fish fed 0% to 6% blood meal. Fasakin, Serwata, and Davies (2005) observed that hybrid tilapia fed 22% blood meal showed reduced feed intake. Similar results were also reported in gilthead sea bream (Martínez-Llorens et al. 2008) and gibel carp (Xue & Cui 2001). If a diet is deficient in essential amino acids, fish may increase feed intake to try and meet these nutrient requirements (Hajen et al. 1993), as happened in the current study. However, when the amino acid deficiency exceeds the minimum value tolerated by the animal, then intake is depressed while dietary amino acid catabolism and lipid retention is enhanced (Aragão et al. 2004). The increase in lipid and energy content in pirarucu whole body with increasing dietary blood meal level agreed with the assumption that dietary protein catabolism increased, possibly caused by an amino acid unbalance from blood meal supplementation alone. In these cases, supplementation of the limiting amino acid either in a purified form or by using a blend of protein-containing ingredients could prevent the reduction of feed intake and nutrient retention. Wang et al. (2008) showed that a blend of poultry by-product meal, meat and bone meal, feather meal, and blood meal was a feasible replacement of fishmeal in malabar grouper, *Epinephelus malabaricus*. El-Haroum and Bureau (2007) compared the bioavailability of lysine in blood meals of various origins to that of L-lysine HCL in trout, and obtained a 7% fishmeal reduction and an increase in lysine content, reducing the need to add synthetic amino acids.

In conclusion, our results indicate that spray-dried blood meal can be moderately supplemented to pirarucu diets for a period of 60 days without affecting growth performance and body. More studies should be carried out to test locally available protein sources, alone or in combination, to be used more extensively as fishmeal substitutes in pirarucu diet.

REFERENCES

Abery, N.W., M.R. Gunasekera, and S.S. De-Silva. 2002. Growth and nutrient utilization of Murry Cod *Maccullochella peelii peelii* (Mitchell) fingerlings fed diets with varying levels of soybean and blood meal. *Aquacult. Res.* 33:279–289.

Allan, G.L., S. Parkinson, M.A. Booth, D.A.J. Stone, S.J. Rowland, J. Frances, and R. Warner-Smith. 2000. Replacement of fishmeal in diets for Australian silver perch, Bidyanus *bidyanus*: I. Digestibility of alternative ingredients. *Aquacult.* 186: 293–130.

Aragão, C., L.E.C. Conceição, D. Martins, I. Ronnestad, E. Gomes, and M.T. Dinis. 2004. A balanced dietary amino acid profile improves amino acid retention in post-larval Senegalese sole (*Solea senegalensis*). *Aquacult.* 233:293–304.

Boorman, K.N. 1992. Protein quality and amino acid utilization in poultry. In *Recent advances in animal nutrition*, edited by P.C. Garnsworthy, W. Haresign, and D.J.A. Cole, 51–70. Boston, MA: Butterwoth-Heinman Ltd.

Bureau, D.P., A.M. Harris, and C.Y. Cho. 1999. Apparent digestibility of rendered animal protein ingredients for rainbow trout (*Oncorhynchus mykiss*). *Aquacult.* 180:345–358.

Castello, L., J.P. Viana, and G. Watkins. 2005. Conditions for community-based management: the pirarucu fishery at the Mimiraua Reserve, Amazon. Paper presented at the XIX annual meeting of the Society for Conservation Biology, held in Brazil in July 2005.

Cavero, B.A.S., D.R. Ituassu, M. Pereira-Filho, R. Roubach, A.M. Bordinhon, F.A.L. Fonseca, and E.A. Ono. 2003a. Use of live food as starter diet in feed training juvenile pirarucu. *Pesq. Agropec. Bras.* 38:1,011–1,015.

Cavero, B.A.S., M. Pereira-Filho, R. Roubach, D.R. Ituassú, A.L. Gandra, and R. Crescêncio. 2003b. Stocking density effect on growth homogeneity of juvenile pirarucu in confined environments *Pesq. Agropec. Bras.* 38:103–107.

Crescencio, R. 2001. Treinamento alimentar de alevinos de pirarucu, *Arapaima gigas* (Cuvier, 1829), utilizando atrativos alimentares (in Portuguese). Master's thesis, Instituto Nacional de Pesquisas da Amazónia–Universidade Federal do Amazonas-UFAM, Manaus, Amazonas, Brazil.

Crossa, M., and J.D.A. Chaves. 2005. The adaptive management of impacts as a strategy for the sustainable use of the pirarucu in the Amazon basin. In *XIX Proc. annual meeting of the Society for Conservation Biology*, edited by M.Â. Marini, 48–49. Ribeiro, Brazil: Universidade de Brasília, Brasília, Society for Conservation Biology.

Davis, D.A., D. Jirsa, and C.R. Arnold. 1995. Evaluation of soybean proteins as replacements for menhaden fishmeal in pratical diets for red drum, *Scianops ocelltus*. *J. World Aquacult. Soc.* 26:48–58.

El-Haroum, E.R., and D.P. Bureau. 2007. Comparison of the bioavailability of lysine in blood meals of various origins to that of L-lysine HCL for rainbow trout (*Oncorhynchus mykiss*). *Aquacult.* 262:402–409.

Fasakin, E.A., R.D. Serwata, and S.J. Davies. 2005. Comparative utilization of rendered animal derived products with or without composite mixture of soybean meal in hybrid tilapia (*Oreochomis niloticus* x *Oreochromis mossambicus*) diets. *Aquacult.* 249:329–338.

Fontenele, O. 1948. Contribuição para o conhecimento da biologia do pirarucu (*Arapaima gigas*, Cuvier), em cativeiro (Actinopterygii, Osteoglossidae) (in Portuguese). *Revista Brasileira de Biologia* 8:445–459.

Gandra, A.L. 2002. Estudo da freqüência alimentar do pirarucu, *Arapaima gigas* (Cuvier, 1829) (in Portuguese). Master's thesis, Universidade Federal do Amazonas, Manaus, Brazil.

Gomes, J.S., and A. Oliva-Teles. 1998. Apparent digestibility coefficients of feedstuffs in seabass (*Dicentrarchus labrax*). *Aquat. Living Resour.* 11:187–191.

Guo, J., Y. Wang, and D. P. Bureau. 2007. Inclusion of rendered anima ingredients as fishmeal substitutes in practical diets for cuneate drum, *Nibea miichthioides*. *Aquacult. Nutr.* 13:81–87.

Hajen, W.E., D.A. Higgs, R.M. Beames, and B.S. Dosanjh. 1993. Digestibility of various feedstuffs by post-juvenile Chinook salmon (Oncorhynchus tshawytcha) in sea water: 2. Measurement of digestibility. *Aquacult.* 112:333–348.

Hayashi, C., G.S.E. Gonçalves, and V.R.B. Furuya. 1999. Utilização de diferentes alimentos durante o treinamento alimentar de alevinos de pintado (*Pseudoplatystoma corruscans*, Agassiz, 1829) (in Portuguese). Proceedings of *Acuicultura 99: Acuicultura en harmonia con el ambiente*. Puerto La Cruz. Venezuela. 258–267.

Kubitza, F. 1995. Preparo de rações e estratégias de alimentação no cultivo intensivo de peixes carnívoros. Proceedings of International Symposium on Fish and Shellfish Nutrition (in Portuguese). *Campos do Jordão*, SP:91–115.

Li, G.Q., and M.V.H. Wilson 1996. Phylogeny of osteoglossomorpha. In: *Interrelations of fishes*, edited by M.L. Stiassny, L.R. Parenti, and G.D. Johnson, 163–174. San Diego, CA: Academic Press.

Luzier, M.J., R.C. Summerfelt, and H.G. Ketola. 1995. Partial replacement of fishmeal with spray-dried blood powder to reduce phosphorus concentrations in diets for juvenile rainbow trout, *Oncorhynchus mykiss* (Walbaum). *Aquacult. Res.* 26:577–587.

Martínez-Llorens, S., A.T. Vidal, A.V. Moñino, J.G. Ader, M.P. Torres, and J.M. Cerdá. 2008. Blood and haemoglobin meal as protein sources in diets for gilthead sea bream (*Sparus aurata*): effects on growth, nutritive efficiency and fillet sensory differences. *Aquacult. Res. 39*:1,028–1,037.

Martins, S.N., and E.C. Guzman. 1994. Effect of drying method of bovine blood on the performance of growing diets for tambaqui (*Colossoma macropomum*, Cuvier 1818) in experimental culture tanks. *Aquacult.* 124:335–341.

Millamena O.M. 2002. Replacement of fish meal by animal by-product meals in a practical diet for grow-out culture of grouper, *Epinephelus coioides*. *Aquacult.* 204:75–84.

Nelson, J.S. 1994. *Fishes of the world, 3rd ed*. New York: John Wiley and Sons.

Pereira-Filho, M., B.A.S. Cavero, R. Roubach, D.R. Ituassú, A.L. Gandra, and R. Crescêncio. 2003. Cultivo do Pirarucu (*Arapaima gigas*) em viveiro escavado (in Portuguese). *Acta Amazônica* 33:715–718.

Pontes, A.C. 1977. O pirarucu *Arapaima gigas* Cuvier, nos açudes públicos do nordeste brasileiro (in Portuguese). Master's thesis, Universidade Federal do Ceará, Fortaleza, Ceará, Brazil.

Queiroz, H.L., and A.D. Sardinha. 1999. A preservação e o uso suatentado dos pirarucus (*Arapaima gigas*, Osteoglossidae) em mamirauá. In *Estratégias de Manejo para Recursos Pesqueiros na Reserva de Desenvolvimento Sustentável Mamirauá*, edited by H.L. Queiroz and W.G.R. Crampton, Mamirauá, Brazil: MCT-CNPq/Sociedade Civil (in Portuguese).

Saint-Paul, U. 1986. Potential for aquaculture of South American freshwater fishes: A review. *Aquacult.* 54:205–240.

Scorvo-Filho, J.D., N.E.T. Rojas, C.M. Silva, and T. Konoike. 2004. Criação de *Arapaima gigas* (Teleostei Osteoglossidae) em estufa e sistema fechado de circulação de água no Estado de São Paulo (in Portuguese) B. *Instituto de Pesca de São Paulo* 30:161–170.

Silva, E.C.S. 2004. Efeito de protease exógena sobre o aproveitamento da proteína vegetal pelo tucunaré paca, *Cichla sp* (in Portuguese). Thesis, Universidade Federal do Amazonas/Instituto Nacional de pesquisas da Amazônia. Manaus, Amazonas, Brazil.

Souza, M.L., and H.C. Menezes. 2004. Processamentos de amêndoas e torta de castanha-do-Brasil e farinha de mandioca: Parâmetros de qualidade (in Portuguese). *Ciência Tecnologia de Alimentos* 24(1):120–128.

Viana, J.P., J.M.B. Damasceno, L. Castello, W.G.R. 2004. Economic incentives for sustainable community management of fishery resources in the Mamirauá Sustainable Development Reserve, Amazonas, Brazil. In *People in nature: Wildlife conservation in South and Central America*, edited by K.M. Silvius, R.E. Bodmer, and J.M.V. Fragoso, 139–154. West Sussex, England: Columbia University Press.

Xue, M., Y.B. Cui. 2001. Effect of several feeding stimulants on diet preference by juvenile gibel carp (*Carassius auratus gibelio*), fed diets with or without partial replacement of fishmealby meat and bone meal. *Aquacult.* 198:281–292.

Zhou, Z., Z. Ren, H. Zeng, and B. Yao. 2008. Apparent digestibility of various feedstuffs for bluntnose black bream *Megalobrama amblyceohala* Yih. *Aquacult. Nutr.* 14:153–165.

Effect of Density on Growth and Feeding of the Crayfish *Cambarellus montezumae* (Saussure, 1857)

JOSÉ LUIS ARREDONDO-FIGUEROA, ANGÉLICA VÁSQUEZ-GONZÁLEZ, IRENE DE LOS A. BARRIGA-SOSA, CLAUDIA CARMONA-OSALDE, and MIGUEL RODRÍGUEZ-SERNA

Planta Experimental de Producción Acuícola, Departamento de Hidrobiología, División de Ciencias Biológicas y de la Salud, Universidad Autónoma Metropolitana Iztapalapa, D.F., Mexico

The acocil C. montezumae *is a freshwater crayfish endemic to the Central Plateau of Mexico, but, in recent years, the natural population of this species has diminished considerably. In this work was investigated growth performance and feeding of this crayfish reared at high densities. A random block experimental design with two repetitions and three treatments (77, 154 and 231 crayfish/m^2) was carried out. Organisms were fed every third day with 15% of their total biomass of shrimp commercial food with 25% of crude protein, and individual food consumption (IFC) was calculated. Significant differences (ANOVA, P < 0.05) were detected in final weight, absolute increase, relative rate of increase, instantaneous rate of increase, yield and survival rate between the densities, with those reared at a density of 77 organisms/m^2 reaching the largest sizes. There were no-significant differences (P > 0.05) among treatments in terms of initial weight, specific growth rate and feed conversion l rate.*

INTRODUCTION

The acocil, *Cambarellus montezumae* is a freshwater crayfish endemic to the Central Plateau of Mexico (Villalobos-Figueroa 1955). Historically, the acocil has been an important source of animal protein for human populations in the Mexico Basin and was consumed by the indigenous populations around lacustrine water bodies. In recent years, its natural populations have diminished considerably due to loss and modification of their habitats, chronic contamination of the lacustrine environment, overexploitation, and the introduction of aggressive exotic species that have become competitors and predators to the extent that this species is now under special protection.

The development of *C. montezumae* culture technology is thus important for both the conservation of the species and its replacement in local economies and diets.

Several authors have studied the effect of stocking density under rearing conditions in Australian species (Geddes, Smallridge, & Clark 1993; Morrissy, Bird, & Cassels 1995; Whisson 1995), the red crayfish *Procambarus clarkii* (Lutz & Wolters 1986; McClain 1995) and the white crayfish *Pascifastacus leniusculus* (Westman 1973; Mason 1979). However, there have not yet been any density and growth rate studies for *Cambarellus montezumae*.

MATERIALS AND METHODS

Experimental Design

A randomized block experimental design with two replications and three treatments was carried out. Juvenile F1 crayfish with initial individual weights ranging from 7.2 to 17.6 mg were randomly introduced into plastic containers (0.69 m L x 0.38 m W, equivalent to 0.26 m^2 in size), which were kept filled to a constant volume of 30 litres during the entire experiment; the densities tested were 77, 154 and 231 crayfish per square meter. Each container was provided with shelters made of straw and scouring pads of natural fibre. As the crayfish grew, these shelters were replaced by half inch PVC tubing, maintaining a constant ratio of one shelter per crayfish. Constant aeration was supplemented during the experiment. The experimental period duration was of four months.

The juveniles were fed every third day with a ration of 15% of their total biomass with commercial shrimp food (25% of crude protein and 3.5% of lipids). To calculate individual food consumption (IFC), the remains of food from the bottom of the container were collected with a siphon and filtered throughout a 3 μm mesh every third day prior to feeding. The residual food was dried at room temperature. The IFC was calculated as: IFC = SF − CF, where SF = supplied food and CF = consumed food in a dry weight basis. The crayfish were weighed every fifteen days to obtain total weight (W), and feeding rate was adjusted for this information. Dead crayfish were substituted with others of similar weight in order to maintain constant density over time. A 100% water replacement was conducted using clean chlorine-free water every week.

During the experimental period, water quality parameters were maintained within the optimum range for this species (Lee & Wickins 1992). Water temperature fluctuated from 19.7 to 20.7°C; dissolved oxygen from 5.1 to 5.4 mg/L and pH from 7.9 to 8.4. NO_2^- never exceeded 0.11 mg/L; NO_3^- was never more than 1.9 mg/L; and total ammonia nitrogen (TAN) not more than 0.25 mg/L.

After 123 days of culture, all animals were harvested and weighted (W). Specific growth rate (SGR) (% day^{-1}) was calculated as SGR = Ln of final weight-Ln of initial weight/culture period in days × 100 (Hopkins 1992); relative rate of increase (RRI) (g), calculated as RRI = fW − iW/iW × 100 (Ricker 1975); instantaneous rate of increase (IRI) (g) as IRI = LnfW − LniW/t (Gulland 1971); feed conversion ratio (FCR) as FCR = Total quantity of consumed food/total weight (Hopkins 1992); yield in g/m^2; survival (%) and final female: male ratio.

Tendency of central measures, dispersion, coefficient of variation and regression were calculated on an Excel spreadsheet. Analyses of Variance (ANOVA) were done using the statistical package JMP version 7.0 (SAS, Institute Inc. Cary, NC, USA). Tests of homocedasticity of variance were applied by means of a robust equality test of Welch, and the Tukey-Kramer HSD (Zar, 2009).

RESULTS

Table 1 shows the growth performance indicators and survival rate registered in the three rearing densities. Significant differences ($P < 0.05$) between the densities were observed in final W, absolute increase (AI), relative rate of increase (RRI), instantaneous rate of increase (IRI) (g), yield (g/m^2) and survival rate, and not were registered significant differences in the initial weight, specific growth rate (SGR) and feed conversion rate (FCR).

Individual Food Consumption

A strong correlation was observed between the IFC and the rearing time in the three densities. In the density of 77 crayfish/m^2 the formula of the linear correlation was IFC = 8.4329 × −330706 ($R^2 = 0.9205$), in 154, IFC = 10.618 × −417068 ($R^2 = 0.8701$) and in 231 IFC = 4.1199 × −162740 ($R^2 = 0.8235$).

TABLE 1 Growth Performance Indicators in a Comparison of *Cambarellus montezumae*. Culture at Three Densities for 123 Days

Indicators	Density (org./m^2)			±S.E.
	77	154	231	
Initial weight (g)	0.040^a	0.039^a	0.036^a	0.0030
Final weight (g)	0.340^a	0.220^b	0.140^c	0.0116
AI (g)	0.30^a	0.19^b	0.11^c	0.0119
SGR (% day^{-1})	1.88^a	1.74^a	1.51^a	0.5773
RRI (g)	8.9^a	7.2^a	4.1^b	0.6361
IRI (g)	2.2^a	1.9^a	1.5^b	0.0881
FCR	1.06^a	1.09^a	1.15^a	0.3337
Yield (g/m^2)	26.0^a	34.2^b	25.4^a	0.5773
Survival (%)	94.0^a	97.0^a	84.0^b	0.6101
♀:♂	18:2	29:11	53:7	

AI = Absolute increase.
RRI = Relative rate of increase.
IRI = Instantaneous rate of increase.
SGR = Specific growth rate.
FCR = Feed conversion ratio.
±S.E. = Standard Error
Average values in row with different superscripts mean significant differences ($P < 0.05$).

DISCUSSION

Several studies regarding rearing density management in several species of crustacean decapods have been carried out. However, this it is the first investigation of rearing conditions for the crayfish *C. montezumae*. In the present study, high stocking densities were evaluated (77, 154 and 231 crayfish/m^2), and the lower density was found to yield the best growth performance. In general, mortality was low, with the first two stocking rates having mortality rates of 6.0 and 2.6% respectively, and the last with 16%.

Domingues & Alaminos (2008) determined that shelters influence the effects of density in culture of the spider crab *Maja brachydactyla*. Kosak et al. (2000) found that the number of shelters is an important factor that affects the intensive or semi-intensive production of crustaceans. A sufficient number of shelters are necessary when managing high densities, since it is a factor that influences cannibalism, especially as crayfish show aggressive behaviours in general. Domingues & Alaminos (2008) also argue that the effect of rearing density is of vital importance, particularly in species with cannibalistic tendencies, since during the ecdysis process the organism is more vulnerable. However, the mortality of cultured organisms can be increased simply by statistical factors, because a higher number of individuals increase the possibility of encounters. These authors also found that the highest mortality occurs after ecdysis, caused mainly by episodes of cannibalism and by the stress generated by this physiological process. This was confirmed by observations that indicated that it is difficult to find dead crayfish with hard exoskeletons, meaning that the cause of death is associated with the ecdysis process. It is important to note that in the present study, the highest mortality was observed in males, likely because they move around more frequently than females.

Optimal rearing conditions have been reported for other species of crustacean decapods. Rodriguez-Serna et al. (2000) reported an optimum density of 50 crayfish/m^2 in *Procambarus llamasi*. Kozak et al. (2000) demonstrated the importance of shelters to reduce mortality in the white crayfish *Pascifasciatus leniusculus*, and recommended a ratio of one shelter per crayfish. Savolainen, Ruohonen and Railo (2004) reared this same species, and found that densities of 200 and 400 crayfish/m^2 can be considered acceptable, demonstrating their ability to tolerate high density rearing. Carmona-Osalde et al. (2004b) indicated that *P. llamasi* also tolerated high rearing densities, with an optimum density of 64 crayfish/m^2 and a relationship of one male for every two females, and concluded that this species had good potential for aquaculture purposes. Oliveira, Marcal and Anastacido (2008) found that it is possible to maintain juveniles of the red crayfish *Procambarus clarkii* in densities of 40 crayfish/m^2 using commercial food. Based on these studies, it is reasonable to conclude that crayfish in general can tolerate high density rearing conditions, compared to other species of crustacean decapods, offering advantages for cultivation in semi- and intensive systems.

Although densities as high as 400 orgs/m^2 were not achieved in this study, it was possible to maintain 200 organisms/m^2 with low mortality. However, the best growth performance indicators were achieved at a density of 77 organisms/m^2. The specific growth rate (SGR) at this density had a value of 1.88% day^{-1}, which is an intermediate value compared to those reported for the redclaw crayfish *Cherax quadricarinatus* (Rodriguez-Canto et al. 2002). For the crayfish *P. llamasi*, SGR values ranged from 0.59 to 2.72% day^{-1} at lower densities (Rodriguez-Serna et al. 2000; Carmona-Osalde et al. 2004a; Carmona-Osalde et al. 2004b). With regard to the food conversion rate (FCR), higher values have been reported for *P. llamasi* (1.9) (Rodriguez-Serna et al. 2000), compared those obtained in the present study (1.06 at 1.15). However, the survival rate achieved in this investigation was 84–94%, higher than that reported for the red crayfish *P. clarkii* (28-80%) (Oliveira, Marcal & Anastacido 2008), and comparable to those obtained for *P. llamasi* (42.5–90% and 85.3–100%) (Carmona-Osalde et al. 2004a, b). The results of the present study indicate that growth performance indicators and nutrition efficiency of the crayfish *C. montezumae* are similar to those reported for other crayfish, and therefore the species can be considered as a good candidate for aquaculture.

Individual Food Consumption

There is no information available on individual food consumption related to aquaculture practices for any species of decapods. However, to gather information on this topic, Ahvenharju and Ruohonen (2005) used an x-ray technique as a direct method to analyse the individual food consumption of *Pascifasciatus leniusculus*. Although there are indirect methods that allow a precise calculation of the consumption, most of them still present technical difficulties mainly due to managed conditions and crayfish physiological states. For example, during the ecdysis and intermoult phases, it has been shown that the energy costs are higher in the juvenile states than in premature and mature organisms, which has consequences in the efficiency of food use (Rombough 1994). It has been shown that a diet based exclusively on vegetable protein is insufficient to maintain a high rate of growth in the crayfish *P. clarkii* (McClain, Neill & Gatlin 1992; Gutierrez-Yurrita & Montes 2001). Crayfishes do have a great nutritional flexibility because they have omnivorous and detritophagus habits (Brown et al. 1989). The commercial food used in this work is elaborated with a base of fish flour to give a content of 25% of animal protein and has given good results for sustaining survival, growth and reproduction in *C. montezumae*. This commercial food has also been used in the culture of red crayfish *P. clarkii* and *P. llamasi*, and growth was not significantly affected. A linear relationship was observed between the average values of individual food consumption and yield. Total food consumption was considered in 0.1 kg of balanced dry food per one hectare per week of the total biomass of the reared crayfish and in a water temperature of 14 to 30°C (McClain 1995).

The results of the present investigation also showed a linear correlation between individual food consumption (IFC) and culture time, reflected in the yields obtained. The highest food consumption was reached at density of 134 crayfish/m^2, which is an equivalent yield of 342 kg/ha. This is less than the value reported by McClain (1995) with an average temperature of 19°C, although this study managed lower densities (10 to 20 crayfish/m^2).

Recent reports indicate that there is a relationship between lipid content in the diet and the growth rate of white crayfish *P. acutus acutus*. Diets with values over 9% lipids resulted in a decrease in weight gain, therefore diets containing between 0 and 6% of lipids are recommended to maintain a high growth rate (Davis & Robinson 2007). The content of lipids of the shrimp food (3.5%) is well within the recommended limit, so this diet did not present negative effects on crayfish growth.

Previous experiments carried out in the Experimental Plant of Aquatic Production (EPAP) demonstrated that it is possible to feed crayfishes with 15% of the total biomass every third day, adjusting the quantity of food given on a biweekly basis. IFC and time showed a high lineal correlation (82 to 92%). The curves indicated that in each ration, the crayfish takes several days to adjust the IFC. These types of curves allowed more precise adjustments of the rations, avoiding food waste, saving money and reducing the cost production.

REFERENCES

Ahvenharju, T., and K. Ruohonen. 2005. Individual food intake measurement of freshwater crayfish (*Pascifastacus leniusculus* Dana) juveniles. *Aquaculture Research* 13(36):1304–1312.

Brown, E.H., A.E. Robinson, A.E. Clark, and A.L. Lawrence. 1989. Apparent digestible energy coefficients and associative effects in practical diets for red swamp crayfish. *Journal of the World Aquaculture Society* 20:122–126.

Carmona-Osalde, C., M. Rodríguez, M.A. Olvera, and P. Gutierrez. 2004a. Gonadal development, spawning, growth and survival of the crayfish *Procambarus llamasi* at three different water temperatures. *Aquaculture* 232: 305–316.

Carmona-Osalde, C., M. Rodríguez, M.A. Olvera, and P. Gutierrez. 2004b. Effect of density and sex ratio on gonad development and spawning in the crayfish *Procambarus llamasi*. *Aquaculture* 236: 331–339.

Davis, D., and E. Robinson. 2007. Estimation of the dietary lipid requirement level of the white crayfish *Procambarus acutus acutus*. *Journal of the World Aquaculture Society* 17:37–43.

Domingues, P., and J. Alaminos. 2008. Efectos de la densidad de cultivo y de elementos de refugio en el crecimiento y supervivencia de juveniles de centollo, *Maja brachydactyla* (Balss, 1922). *Revista de Biología Marina y Oceanografía* 43(1):121–127.

Geddes, M.C., M. Smallridge, and S. Clark. 1993. The effect of stocking density, food type and shelters on survival and growth of the Australian freshwater crayfish, *Cherax destructor*, in experimental ponds. *Freshwater Crayfish* 9:57–69.

Gulland, J.A., 1971. Ecological aspects of fisheries research. *Advances in Ecological Research* 7:115–176.

Gutiérrez-Yurritia, P.J., and C. Montes. 2001. Bioenergetics of juveniles of red swamp crayfish (*Procambarus clarkii*). Elsevier Science 130:29–38.

Hopkins, K.D. 1992. Reporting fish growth: a review of the basis. *Journal of the World Aquaculture Society* 23: 173–179.

Kosak, P., J. Kajtman, J. Kouñil, and T. Policar. 2000. The effect of crayfish density and shelter number on the daily activity of signal crayfish. *Freshwater Crayfish* 13:457–462.

Lee, D.O.C., and J.F. Wickins. 1992. Crustacean farming. Halsted Press; John Wiley & Sons, Inc., New York, USA, 236 p.

Lutz, R.S., and W.R. Wolters. 1986. The effect of five stocking densities on the growth and yield of red swamp crayfish *Procambarus clarkii*. *Journal of the World Aquaculture Society* 17:33–36.

Mason, J.C. 1979. Effects of temperature, photoperiod, substrate and shelter on survival, growth and biomass accumulation of juvenile *Procambarus leniusculus*. *Freshwater Crayfish* 4:73–82.

McClain, W.R., W.H. Neill, and D.M. Gatlin. 1992. Nutrient profiles of green and decomposed rice-forages and their utilization by juvenile crayfish (*Procambarus clarkii*). *Aquaculture* 101:251–265.

McClain, W.R. 1995. Effects of population density and feeding rate on growth and feed consumption of Red Swamp Crayfish *Procambarus clarkii*. *Journal of the World Aquaculture Society* 26(1):14–23.

Morrissy, N.M., C. Bird, and G. Cassells. 1995. Density-dependent growth of cultured marron, *Cherax tenuimaus* (Smith, 1912). *Freshwater Crayfish* 10:560–568.

Oliveira, R., A. Marcal, and P. Anastacido. 2008. Effects of density on growth and survival of juvenile Red Swamp Crayfish, *Procambarus ÿclarkii* (Girard), reared under laboratory conditions. *Aquaculture Research* 39:577–586.

Ricker, W.E. 1975. Computation and interpretation of biological statistics of fish populations. *Fisheries Research Board of Canada Bulletin*,191:382 p.

Rodríguez-Canto, A., J.L. Arredondo-Figueroa, J.T. Ponce-Palafox, and D. Rouse. 2002. Growth characteristics of the Australian Redclaw Crayfish, *Cherax quadricarinatus*, cultured in an indoor recirculating system. *Journal of Applied Aquaculture* 12 (3):59–64.

Rodriguez-Serna, M., C. Carmona, M.A. Olvera, and J.L. Arredondo. 2000. Fecundity, egg development and growth of juvenile crayfish *Procambarus (Austrocambarus) llamasi* (Villalobos 1955) under laboratory conditions. *Aquaculture Research* 31: 173–179.

Rombough, P.J. 1994. Energy partitioning during fish development: additive or compensatory allocation of energy to support growth? *Functional Ecology* 8:605–610.

Savolainen, R., K. Ruohonen, and E. Railo. 2004. Effect of stocking density on growth, survival and cheliped injuries of stage 2 juvenile signal crayfish *Pascifastacus leniusculus* Dana. *Aquaculture* 231: 237–248.

Villalobos-Figueroa, A. 1955. Cambarinos de la fauna mexicana, Crustácea: Decapoda. Tesis de Doctorado, Facultad de Ciencias, Universidad Nacional Autónoma de México, p. 290.

Westman, K. 1973. Cultivation of the American crayfish *Pascifastacus leniusculus*. *Freshwater Crayfish* 1:221–232.

Whisson, G.J. 1995. Growth and survival as a function of density for marron *Cherax tenuimanus* (Smith) stocked in a recirculating system. *Freshwater Crayfish* 10:630–637.

Zar, H.J. 2009. *Biostatistical analysis*. 5th Edition, London: Prentice Hall, p. 944.

Growth of Juvenile Crayfish *Procambarus llamasi* (Villalobos 1955) Fed Different Farm and Aquaculture Commercial Foods

M. RODRIGUEZ-SERNA, C. CARMONA-OSALDE,
and J. L. ARREDONDO-FIGUEROA

Universidad Autónoma Metropolitana Iztapalapa (UAM-I), División de Ciencias Biológicas y de la Salud, Departamento de Hidrobiología, Planta Experimental de Producción Acuícola, Delegación Iztapalapa, Distrito Federal, México

In this study, a range of commercially available animal and fish feeds were tested in the production of juveniles of the crayfish Procambarus llamasi, *an endemic species of the Yucatan Peninsula, Mexico. A randomized block design was implemented to observe the effect of commercial shrimp (Crude Protein, CP = 38%), trout (CP = 43.2%), tilapia (CP = 31.8%), rabbit (CP = 16.1%), turkey (CP = 18%), and pig (CP = 15.2%) diets in a recirculating aquaculture system. Shrimp food showed the best results in terms of growth performance, while lowest growth values (P<0.05) were obtained with farm animal diets. Trout and tilapia diets were intermediate. Pig feed was the most cost-effective at US$ 0.44 per kg of crayfish growth. Where no feeds specifically for crayfish are available, terrestrial animal feeds have the advantage of low price, ready availability in the market, and acceptable performance.*

The investigators wish to thank the Conacyt-Gobierno del estado de Quintana Roo for its financial support of the project as well as the Provi-Purina Co. of Yucatan for the donation of the foods. This experiment supports the technological transfer of the *Procambarus (A.) llamasi* in the State of Quintana Roo.

INTRODUCTION

The crayfish, or *acocil*, a common name given in Mexico for members of the family Cambaridae, has become a potential resource for aquaculture in Mexico. In spite of the numerous crayfish species in the world, only a few are cultured and there is little or scarce biological information for most of the other species (Huner 1991;Avault 1992; McClain & Romaire 2009).

In the Yucatan Peninsula there exists an endemic species of crayfish, *Procambarus (Austrocambarus) llamasi*, that could be of interest for regional aquaculture (Rodríguez-Serna et al. 2002). Rodriguez-Serna and colleagues (2000) showed that this species has a growth rate similar to that of the red swamp crayfish, *Procambarus clarkii* (Huner1988) reaching commercial size (12 to 14 cm) in only four to six months. In addition, this species can tolerate high stocking densities without affecting their growth rate (80–100 org. m^{-2}), reproduce naturally throughout the year (with three annual peaks), can be reproduced under controlled conditions (Carmona-Osalde, Rodríguez-Serna, & Olvera-Novoa 2002; Carmona-Osalde et al. 2004a, 2004b; Carmona-Osalde, Rodríguez-Serna, & Olvera-Novoa 2005), and are easily adapted to formulated food (Rodríguez-Serna, Carmona-Osalde, & Olvera-Novoa 1998).

Most of the crayfishes have been classified as detritivorous (Avault, Romaire, & Miltner 1981; McLarney 1984), herbivorous (Iheu & Bernardo 1993), or omnivorous/opportunistic feeders (D'Abramo & Robinson 1989), and in natural conditions their stomach contents feature up to 80% detritus and vegetable matter (Gutiérrez-Yurritaeta 1997) with the balance comprised of small molluscs, nematodes, and insects (Loya-Javellana, Fielder, & Thorne 1993).

According to Brown (1995), plant and detrital materials contribute dietary carotenoids, other pigments, and energy, but at least some species show a preference for diets based on animal meals (Rodríguez-Serna, Carmona-Osalde, & Olvera-Novoa 1998; Guan & Wiles 1998), and are highly efficient assimilators of such materials. Jones and colleagues (1997), for example, reported efficiency of assimilation of animal matter up to 85% for *Cherax tenuimanus* and 88%-96% in *Cherax destructor*.

In Mexico, there are no commercial crayfish diets for these crustaceans. The aquaculture of these crustaceans in extensive and semi-intensive systems using low-quality feeds could reduce production costs. The target of the present work was to probe the effect of different commercially available animal foods used for terrestrial and aquatic species like rabbit, pig, poultry, rainbow trout, tilapia, and shrimp foods in the semi-intensive production of the crayfish *P. llamasi*.

MATERIALS AND METHODS

Crayfish were grown in a closed recirculating system with 288 individual containers to avoid cannibalism (Figure 1). Temperature was kept constant at 26 ± 1°C, and flow of water was maintained in 4 L h^{-1}.

Juvenile crayfish of an initial average weight of 1.16 g ± 0.7 g were stocked into the individual units in a completely randomized block experimental design, with three replicates per treatment. Six commercial diets (Purina®) available in local stores were compared, three of which were designed for aquaculture (shrimp, tilapia, and rainbow trout) and the other three for land animal husbandry (rabbit, pig, and turkey). The crayfish were fed twice daily (08:00 and 16:00 h) at 10% of total body weight. All the diets were given on plates to avoid losses. Remaining feed and faeces were siphoned off daily. Results of proximate diet analysis done according to standard methods (Association of Official Analytical Chemists 1984) are presented in Table 1. Growth performance indicators used in this experiment were calculated according to Olvera and colleagues (1990):

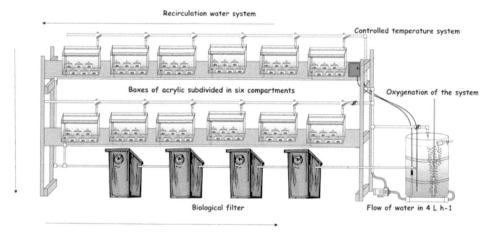

FIGURE 1 Schematic drawing of system used in the experiment.

TABLE 1 Proximal Composition (%) of the Commercial Diets Used in this Experiment

Parameters	Shrimp	Rabbit	Turkey	Tilapia	Trout	Pig
Moisture	6.43	6.42	6.44	8.56	8.55	6.40
Protein	38.	16.1	18.0	31.8	43.2	15.2
Lipids	9.3	1.4	3.8	1.7	8.3	2.8
Fiber	3.3	53.5	51.1	4.9	4.0	53.5
Ash	11.3	16.82	14.3	6.1	9.9	15.1
*Free Extract of nitrogen	31.6	5.8	6.3	46.9	26.0	7.1

*F.E.N. = (100% - [Humidity + Protein + Lipids + Fiber + Ash]).

Survival [% = 100 ((initial number - final number) / total number)]
Absolute weight gain (**Gw** mg day^{-1}) = 100 (weight initial − weight final / time)
Relative weight gain (**Gw** %) = 100 ((weight final − weight initial) / weight initial)
Specific growth rate (**SGR** % day^{-1}) = 100 ((Ln final weight − Lnweight initial)/time)
Individual food consumed (**IFC** g day^{-1}) = 1000 (feed initial − feed final/time)
Feed conversion ratio (**FCR**) = IFC / Gw

Results were compared by means of one-way analysis of variance (ANOVA), and the differences were contrasted between the treatments using Tukey's Multiple Range Test (Zar 1984). The statistical analysis was made with Statgraphics version 4.0.

RESULTS

Survival was 100% in all the treatments, indicating good acceptance of all of the foods. Maximum growth rate was obtained with the diets containing more than 30% crude protein. Non-significant differences were observed among diets except for shrimp food. The results of growth and composition of food are presented in Table 2. Although the rainbow trout food contained high levels of protein, it did not elicit better growth performance compared

TABLE 2 Growth Performance Indicators in the Six Commercial Diets Used in this Experiment and its Cost per Kilo

Variables[1]	Shrimp	Rabbit	Turkey	Tilapia	Trout	Pig	± E.S.
Initial weight (g)	1.25^a	1.13^a	1.18^a	1.12^a	1.13^a	1.15^a	0.0394
Final weight (g)	4.69^c	2.48^{ab}	2.14^a	2.83^b	2.91^b	2.55^{ab}	0.2120
Gain weight (g day^{-1})	3.44^c	1.32^{ab}	0.95^a	1.71^b	1.78^b	1.40^{ab}	0.2111
Gain weight (%)	278.9^c	122.5^{ab}	86.2^a	156.2^b	160.9^b	127.3^{ab}	18.577
Specific Growth Rate (% day^{-1})	1.44^c	0.07^{ab}	0.62^a	1.01^b	1.02^b	0.86^b	0.0762
Individual Food Consumed (g day^{-1})	2.49^b	1.69^a	1.75^a	1.74^a	1.82^a	1.71^a	0.0759
Feed Conversion Ratio (FCR)	1.01^a	1.28^a	4.5^b	1.15^a	1.17^a	1.46^a	2.2283
Cost of food ($US Dollar/kg)	1.12	0.40	0.43	0.74	0.97	0.36	
Cost Kg of growth in crayfish ($US dollar)	0.81	0.51	0.79	0.75	0.99	0.44	

[1]Superscript with different letters means significantly different (P>0.05).
± E.S. = Error standard.

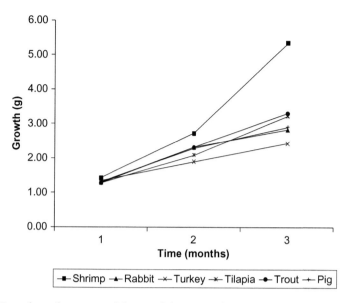

FIGURE 2 Growth performance of the crayfish *Procambarus (A.) llamasi* during the experimental period.

with the other foods. The best growth ($p \leq 0.05$) was obtained with the diet formulated for shrimp (Figure 2).

Likewise, food consumption values were similar among the diets except for shrimp food, where the final value was higher. FCRs were not different between diets, except for turkey food, which had a very high value. The best FCRs ($p \leq 0.05$) were obtained with the diets for aquatic organisms in comparison with the land animal diets (Table 2).

Comparison of the economic performance of foods (Table 2) shows that rabbit and pig foods are the most cost-effective alternatives up to the size of crayfish produced in this study (about one third of marketable size). Shrimp is the more expensive food item used with a price per kilo of 1.12 dollars. The cheapest commercial food was pig with a price per kilo of 0.36 dollars (Table 2).

DISCUSSION

Among the foods tested here, shrimp food provides the best performance. This food appears to come closer to covering the nutritional requirements of crayfishes and has good stability in the water, reducing nutrient leaching and subsequent water quality problems. However, it is important to note that the diets for farm husbandry, especially rabbit, showed good acceptance and good results in the event one does not have shrimp food. These results are very similar to those obtained previously by Rodríguez-Serna, Carmona-Osalde, & Olvera-Novoa (1998).

Although the protein content of the rainbow trout and tilapia diets were superior or similar to that of shrimp food, these did not produce growth rates similar to that of shrimp food, possibly due to lower stability in the water, as in the case of the foods for farm husbandry. The loss of nutrients through leaching is a problem often encountered with slow-eating crustaceans (Tacon & Silva 1997).

The fact that high fiber content in the land animal feeds had no obvious detrimental effect on growth and survival (compared to the trout and tilapia diets) reflects the generally low nutritional quality of crayfish diets in the wild (Hessen & Skurdal, 1986). In crayfish, intestinal bacteria contribute 18 of the 20 necessary free amino acids for healthy growth (Syvokiené & Mickéniené 1993).

The diet with the worst performance in terms of growth and feed utilization was the turkey food. Although foods with a low relationship of C:N are generally good for crayfish, the main problem seems to have been the particle size of the different ingredients that give it poor stability in the water and a quick disintegration.

Crayfish can survive on plants and detritus, but this food does not produce maximum growth (Verhoef, Jones, & Austin 1998; García-Ulloa et al. 2003). Although this material represents the biggest percentage in the stomach content under natural conditions, crayfishes should not be considered strictly detritivorous (Brown et al. 1989; Reigh, Braden, & Craig1990; Huner 1997). Their diet requires protein contents between 25% and 30% to improve growth rates in short periods of time in recirculation aquaculture systems (Huner & Meyers 1979; McLarney 1984; Huner 1991; Verhoef, Jones, & Austin1998; Jover et al. 1999; Rodríguez-Serna et al. 2000).

In this study, *P. llamasi* showed a great capacity for acceptance and adaptation to any kind of food, making them an attractive culture species for farmers in areas where prepared diets are scarce and/or expensive. In fact, in terms of cost-effectiveness, pig and rabbit foods were superior to fish and shrimp foods, although whether or not this would be reflected in increased profitability of a system producing market-sized crayfish (about 15g) will require further research.

REFERENCES

Association of Official Analytical Chemists (AOAC). 1990. *Official methods of analysis of the Association of Analytical Chemist, 15^{tb}ed*. Arlington, VA: AOAC.

Avault, J.W. Jr., R.P. Romaire, and M.R. Miltner. 1981. Feeds and forages for red swamp crawfish, *Procambarus clarkii*: 15 years research at the Louisiana State University reviewed. *Freshwater Crayfish* 5:362–369.

Avault, J.W. Jr. 1992. A review of world crustacean aquaculture part two. *Aquaculture Magazine*, July/August 18(4): 83–92.

Brown, P.B., E.H. Robinson, A.E. Clark, and L. Lawrence.1989. Apparent digestible energy coefficients and associative effects in practical diets for red swamp crayfish. *J. World Aquacult. Soc.* 20(3): 122–126.

Brown, P.B. 1995.A review of nutritional research with crayfish.*J. Shellfish Res.*14(2): 561–568.

Carmona-Osalde, C., M. Rodríguez-Serna, and M.A. Olvera-Novoa. 2002. The influence of the absence of light on the onset of first maturity and egg laying in the crayfish *Procambarus* (*Austrocambarus*) *llamasi* (Villalobos 1955). *Aquaculture* 212(1–4): 289–298.

Carmona-Osalde, C., M. Rodríguez-Serna, M.A. Olvera-Novoa, and P.J. Gutiérrez-Yurrita. 2004a. Gonadal development, spawning, growth and survival of the crayfish *Procambarus* (*Autrocambarus*) *llamasi* on the three different water temperatures. *Aquaculture* 232:305–316.

Carmona-Osalde, C., M. Rodríguez-Serna, M.A. Olvera-Novoa, and P.J. Gutiérrez-Yurrita. 2004b. Effect of density and sex ratio on gonad development and spawning in the crayfish *Procambarus llamasi*. *Aquaculture* 236:331–339.

Carmona-Osalde, C., M. Rodríguez-Serna, and M.A. Olvera-Novoa. 2005. Effect of the protein-lipids ratio on growth and maturation of the crayfish *Procambarus* (*Austrocambarus*) *llamasi*. *Aquaculture* 250:692–699.

D'Abramo, L.R., and H. Robinson.1989. Nutrition of crayfish. *Rev. Aquat. Sci.* 1(4): 711–728.

Garcia-Ulloa, G.M., H.M. López-Chavarín, H. Rodríguez-González, and H. Villareal-Colmenares. 2003. Growth of redclaw crayfish *Cherax quadricarinatus* (von Martens 1868) (Decapoda: Parastacidae) juveniles fed isoproteic diets with partial or total substitution of fish meal by soya bean meal: Preliminary study. *Aquacult. Nutr.* 9:25–31.

Goddard, J.S. 1988. Food and feeding. In *Freshwater crayfish, biology, management and exploitation*, Eds. D.M. Holdich andR.S. Lowery, 145–166. London: Croom Helm Press.

Guan, R., and P.R. Wiles.1998. Feeding ecology of the signal crayfish *Pascifastacusleniusculus* in a British lowland river. *Aquaculture* 169:177–193.

Gutiérrez-Yurrita, P.J., S. Gorka, M.A. Bravo, A. Baltanás, and C. Montes. 1997. Diet of the red swamp crayfish *Procambarus clarkii* in natural ecosystems of the Doñana National Park freshwater marsh (Spain). *J. Crustacean Biol.* 8(1): 120–127.

Hessen, D.O., and J. Skurdal.1986. Analysis of food utilized by the crayfish *Astacusastacus* in lake Steinsfjorden S.E. Norway. *Freshwater Crayfish* 5: 187–193.

Huner, J.V., and S.P. Meyers.1979. Dietary protein requirements of the red crawfish, *Procambarus clarkii* (Girard) (Decapoda: Cambaridae), grown in a closed system. *Proc. World Maricult. Soc.*10:751–760.

Huner, J.V., and J.E. Barr. 1984. *Red swamp crayfish: Biology and exploitation*. Baton Rouge, LA: Louisiana Sea Grant College Program, Center for Wetland Resources, Louisiana State University.

Huner, J.V. 1997. Just how important is detritus as crayfish food? *Crayfish News* 19(4): 10.

Huner, J.V. 1988. *Procambarus* in North America and elsewhere. In *Freshwater crayfish: Biology, management and exploitation*, Eds. D.M. Holdich and R.S. Lowery, 239–261. London: Croom Helm Press.

Huner, J.V. 1991.Aquaculture of freshwater crayfish. In *Production of aquatic animals: Crustaceans, Mollusks, Amphibians, and Reptiles, World Animal Science*, Ed. C.E. Nash, 45–66. Amsterdam, Netherlands: Elsevier Science.

Iheu, M., and J.M. Bernardo. 1993. Experimental evaluation of food preferences of red swamp crayfish, *Procambarusclarkii*: vegetable *versus* animal. *Freshwater Crayfish* 9:359–364.

Jones, D.A., M. Kumlu, L. Le-Vay, and D.J. Fletcher. 1997. The digestive physiology of herbivorous, omnivorous and carnivorous crustacean larvae: a review. *Aquaculture* 155:285–295.

Jover, M., J. Fernández-Carmona, M.C. del Río, and M. Soler. 1999. Effect of feeding cooked-extruded diets, containing different levels of protein, lipid and carbohydrate on growth of red swamp crayfish (*Procambarusclarkii*). *Aquaculture* 178:127–137.

Loya-Javellana, G.N., D.R. Fielder, and M.J. Thorne. 1993. Food choice by free living stages of the tropical crayfish *Cherax quadricarinatus* (Parastacidae: Decapoda). *Aquaculture* 29:261–278.

McLarney, W. 1984. *The freshwater aquaculture book*. Vancouver, BC, Canada: Hartley & Marks Press.

McClain, W.R., and R.P. Romaire. 2009. Contribution of different food supplements to growth and production of res swamp crayfish. *Aquaculture* 249:93–98.

Olvera, N.M.A., G.S. Campos, G.M. Sabido, and P.C.A. Martínez. 1990. The use of alfalfa leaf protein concentrates as a protein source in diets for tilapia (*Oreochromis mossambicus*). *Aquaculture* 90:291–302.

Reigh, R.C., S.L. Braden, and R.J. Craig,1990. Apparent digestibility coefficients for common feedstuffs in formulated diets for red swamp crayfish, *Procambarus clarkii*. *Aquaculture* 84:321–334.

Rodríguez-Serna, M., C. Carmona-Osalde, and M.A. Olvera-Novoa. 1998. Growth and survival of juvenile crayfish *Procambarus (A.) llamasi* fed with fresh or artificial diets. Proceedings of the 12th International Symposium of Astacology (IAA), Haus St. Ulrich, Augsburg, Germany.

Rodríguez-Serna, M., C. Carmona-Osalde, M.A. Olvera-Novoa, and J.L. Arredondo-Figueroa. 2000. Fecundity, egg development and growth under two densities of juvenile crayfish *Procambarus (Austrocambarus) llamasi* (Villalobos 1955) under laboratory conditions. *Aquacult. Res.* 31(2): 173–180.

Rodríguez-Serna, M., C. Carmona-Osalde, J.L. Arredondo-Figueroa, and M.A. Olvera-Novoa. 2002. Distribución geográfica actual de *Procambarus (Austrocambarus) llamasi* (Cambaridae: *Procambarus*) en la Península de Yucatán. *Hidrobiológica* 12(1): 1–5.

Syvokiené, J., and L. Mickéniené.1993. The activity of gut bacteria of the crayfish, *Pascifastacus leniusculus* (Dana), in producing essential free amino acids. *Freshwater Crayfish* 9:235–240.

Verhoef, G.D., P.P. Jones, and C.M. Austin. 1998. A comparison of natural and artificial diets for juveniles of Australian freshwater crayfish *Cherax destructor*. *J. World Aquacult. Soc.* 29(2): 243–248.

Tacon, A., and S.S. de Silva, 1997. Feed preparation and feed management strategies within semi-intensive fish farming systems in the tropics. *Aquaculture* 151(1–4): 379–404.

Zar, J.H., 1984. *Biostatistical Analysis*. Upper Saddle River, NJ: Prentice-Hall.

Survival of Diploid and Triploid *Rhamdia quelen* Juveniles Under Different Oxygen Concentrations

LUCIANO AUGUSTO WEISS and EVOY ZANIBONI-FILHO

Laboratório de Biologia e Cultivo de Peixes de Água Doce (LAPAD), Departamento de Aqüicultura, Centro de Ciências Agrárias (CCA), Federal University of Santa Catarina (UFSC), Florianópolis, Brazil

The objective of this study was to determinate the lethal concentration of dissolved oxygen (DO) over 96 hours of exposure (LC50–96h) for diploid and triploid jundia Rhamdia quelen *juveniles. Diploid and triploid fish weighing approximately 4 g were subjected to DO concentrations varying between 0.4 and 1.3 mg O_2 L^{-1}; water temperature was maintained at 27°C and pH at 6.3. The LC50–96h for diploids of* Rhamdia quelen *was 0.535 mg O_2 L^{-1}, while the value obtained for triploids was 6% greater. These results demonstrated that triploids of* Rhamdia quelen *juveniles have greater sensitivity to hypoxia compared to diploids.*

INTRODUCTION

Dissolved oxygen (DO) is considered one of the main factors that limit productivity in fish culture systems (Boyd & Watten 1989, Boyd 1990). Low oxygen concentrations (hypoxia) reduce fish growth (Chabot & Dutil 1999; Karim, Sekine, & Ukita 2002; Wu 2002; Wilhelm et al. 2005; Maffezzolli & Nuñer 2006), induce damage to fish gill tissue (Wu 2002), reduce food intake and food conversion (Bergheim et al. 2006; Maffezzolli & Nuñer 2006), and affect fish survival (Braun et al. 2006) and reproductive capacity (Karim, Sekine, & Ukita 2002).

Low DO concentrations induce anaerobic metabolism in fish (Cooper et al. 2002; Wu 2002), which results in toxic effects on the organisms. This condition not only causes stress, but can also increase mortality (Madenjian, Rogers, & Fast 1987). The lethal concentration that results in the death of approximately 50% of the fish population (LC50) is used to estimate mortality for short period tests, generating a safe range for this toxic concentration for both medium and long-term tests (Greenburg 1992).

The jundia *Rhamdia quelen* (Quoy & Gaimard 1824; Siluriformes: Heptapteridae) is a catfish with a wide geographical distribution that can be found from central regions of Argentina to southern Mexico (Silfvergrip 1996). In the natural environment, this species presents nocturnal behavior and lives in deep and lentic areas of rivers, suggesting that it can support significant fluctuations in DO (Gomes et al. 2000). In recent years, the attention given to this species by fish farmers in southern Brazil has increased due to it being relatively easy to handle with satisfactory growth in low water temperatures (Carneiro et al. 2002, Fracalossi et al. 2004) and the excellent acceptance of this catfish by the consumer market (Marchioro & Baldisserotto 1999, Barcellos et al. 2001).

However, there are some difficulties in large-scale operations of jundia, including early sexual maturation and a heterogeneous growth ratio (Fracalossi, Zaniboni-Filho, & Meurer 2002). Males present a lower weight gain than females, even though both sexes utilize metabolic energy for the production of gametes before reaching commercial size (Fracalossi et al. 2004). Chromosomic manipulation to generate triploids is being utilized as a method to eliminate sexual maturation, which also results in other advantages. Triploidy in some species of Siluriformes can guarantee higher weight gain (Wolters, Chrisman, & Libey 1982; Fast et al. 1995), improved feed conversion (Wolters, Chrisman, & Libey 1982), and the desired reproductive sterility (Tiwary, Kirubagaran, & Ray 2000).

Triploid fish also exhibit bigger cells and a decreased total cell number, especially in the central nervous system and sensory organs, suggesting different perception, behavior, and physiological responses compared to diploids (Small & Benfey 1987; Aliah et al. 1990; Aliah et al. 1991; Kavumpurath & Pandian 1992; Benfey 1999). The objective of this study was to determinate the lethal concentration of DO over 96 hours of exposure (LC50–96h) for diploid and triploid jundia *Rhamdia quelen* juveniles.

MATERIALS AND METHODS

Fish

This study was conducted at the facilities of Federal University of Santa Catarina (27° 43' 44"S/48° 30' 32"W). Wild jundia broodstock from the Upper Uruguay River were used to produce juveniles. Hormonal induction of two couples produced gametes that were pooled and then separated to produce diploids and triploids.

Triploids were generated using a pressure shock of 5,000 psi for 5 minutes that was applied 4 minutes after fertilization, according to Huergo and Zaniboni-Filho (2006). Diploid and triploid eggs were maintained separately in conical-cylinder incubators with constant water flow. After hatching, larvae were kept separately in 1,000 L tanks connected with a closed water circulation system. During this period, fish larvae were maintained in lightly salted water (3‰ of NaCl) and fed on a diet of *Artemia* sp. *nauplii*.

All fish obtained from the pressure shock treatment were analyzed to confirm the ploidy type using the method described by Phillips and colleauges (1986), with modifications (Huergo & Zaniboni-Filho 2006). The method involves utilization of $AgNO_3$ to pigment cellular nuclei, followed by quantification of nuclei, with three nuclei in a single cell indicating a triploid individual. The treatment used was efficient to induce 100% triploids in the treated group.

Reared Conditions

Each group of 10 fish was placed in a 30 L black plastic aquarium that was distributed in a completely randomized design, with three replicates. Diploids (body weight 3.91 ± 0.98 g; length 9.55 ± 0.78 cm, mean ± S.D.) and triploids (4.28 ± 1.27 g; 9.50 ± 0.92 cm) were stocked separately using similar biomass for each aquarium. Four oxygen regimes were applied simultaneously (0.4, 0.7, 1.0, and 1.3 mg O_2 L^{-1}) for each ploidy treatment during the four-day experiment (96 h).

The chosen DO concentrations were based on a previous study performed by Braun et al. (2006) that determined the LC50–96h for diploid juveniles of jundia. To obtain the desired DO concentrations, each aquarium was maintained with no water recirculation, while compressed air and/or pure nitrogen were added if necessary to maintain those concentrations. Gaseous exchange interference between atmospheric air and water was reduced by completely covering the aquarium water surface with plastic.

DO and temperature were measured every four hours in each aquarium for maintenance of selected treatment levels. All measurements of water temperature and DO were performed using YSI 55 equipment (Yellow Springs, OH). Every eight hours, pH was measured using a pH meter (YSI 60, Yellow Springs, OH). Alkalinity, ionized ammonia (NH_3), and nitrite were determined using a colorimetric kit (Alpha Tecnoquimica), the indofenol method (Koroleff 1975), and the method described by Golterman, Clymo, and Ohnstad (1978), respectively, at both the start and end of the experiment. Fish mortality was quantified every four hours.

Statistical Analysis

Mean LC50 for dissolved oxygen was calculated using the Probit Program developed by the U.S. Environmental Protection Agency (EPA 1984). Data for water quality and mortality were tested for significant differences by analysis of variance (ANOVA) followed by the Tukey test ($P < 0.05$) using GraphPad InStat.

RESULTS AND DISCUSSION

Water quality was similar among treatments, except for DO (Table 1). Alkalinity was measured at 27.50 ± 2.89 mg $CaCO_3$ L^{-1} (mean $\pm$ S.D.).

Survival of fish was 100% in both diploids and triploids exposed to the highest DO level for 96 hours. Accumulated average mortality was similar between both ploidys, except for the treatment of 0.70 mg O_2 L^{-1}, for which triploids exhibited a greater mortality than diploid individuals (Figure 1).

The viability of triploid fish for commercial aquaculture depends on the potential advantages when compared with diploids, mainly related to the physiological adaptability and immune resistance of these fish. The similarity of the immune responses between ploidys is well known, and it is also known that the lesser number of leukocytes found in triploids is compensated for by a greater fish size and higher cellular activity (Budiño et al. 2006). Comparative studies that evaluate the survival of diploids and triploids reveal conflicting results and variability between different fish species (Kerbya et al. 2002).

Similar fish behavior was observed for both ploidys and treatments, except under the higher oxygen concentration. Under DO concentration at 1.0 mg O_2 L^{-1} and below, triploid and diploid fish lost scale color, swam closely to the surface, increased operculum movement, lost equilibrium, exhibited convulsions, and some individuals died. Dead fish presented an extreme opening of the operculum. Considering the mortality rate in all

TABLE 1 Mean Values ($\pm$S.D.) for Dissolved Oxygen, Temperature, pH, Unionized Ammonia, and Nitrite Observed During the Experiments for Both Ploidys of Jundia *Rhamdia quelen*

Ploidy	DO (mg L^{-1})	Temperature (°C)	pH	NH_3 (mg L^{-1})	NO_2^- (mg L^{-1})
2n	0.44 ± 0.04 a	27.02 ± 0.07 a	6.40 ± 0.06 a	0.015 ± 0.015 a	0.079 ± 0.089 a
3n	0.44 ± 0.04 a	27.05 ± 0.04 a	6.38 ± 0.07 a	0.014 ± 0.014 a	0.079 ± 0.089 a
2n	0.70 ± 0.05 b	27.04 ± 0.07 a	6.33 ± 0.08 a	0.017 ± 0.017 a	0.074 ± 0.084 a
3n	0.70 ± 0.05 b	27.01 ± 0.06 a	6.34 ± 0.08 a	0.017 ± 0.017 a	0.076 ± 0.087 a
2n	1.01 ± 0.08 c	27.01 ± 0.07 a	6.26 ± 0.04 a	0.021 ± 0.022 a	0.096 ± 0.110 a
3n	1.00 ± 0.08 c	27.03 ± 0.05 a	6.25 ± 0.07 a	0.021 ± 0,022 a	0.096 ± 0.110 a
2n	1.30 ± 0.07 d	27.05 ± 0.06 a	6.32 ± 0.08 a	0.019 ± 0.020 a	0.128 ± 0.147 a
3n	1.33 ± 0.07 d	27.01 ± 0.07 a	6.33 ± 0.05 a	0.017 ± 0.017 a	0.122 ± 0.139 a

Different letters in a column indicate a significant difference (Tukey, $P < 0.05$).

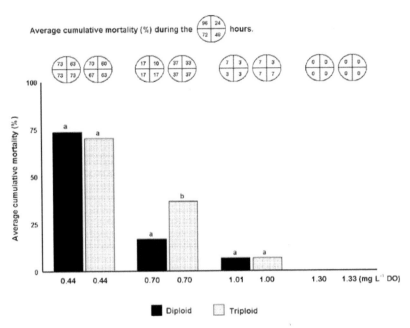

FIGURE 1 Cumulative mortality average (%) for diploids and triploids of jundia *Rhamdia quelen* exposed to different DO concentration during 96 h. Different letters in the same treatment indicate a significant difference (Tukey, $P < 0.05$). Circular graphs above each bar represent the cumulative mortality average (clockwise) for each treatment after each 24 h interval of exposition.

treatments collectively after 96 h, triploids exhibited a 17.2% higher mortality than diploids (Figure 2).

The greater mortality of triploids under hypoxia conditions could be a result of a slow physiological adaptation to this stressor. Triploids have been shown to present a reduction in cell number of the central nervous system and a decrease in hormone levels, demonstrating a slower reaction to environment stimulation when compared to diploids (Benfey 1999).

These differences in the number and size of cells can also affect the behavior and physiology of triploids, producing fish that are less sensitive to sound and light (Aliah et al. 1990), and fish that are relatively less aggressive (Kavumpurath & Pandian 1992).

The increase in exposure time raised the LC50 values similarly for both ploidys, increasing the expected value by approximately 10% from 24 to 96 h of exposure (Table 2). The LC50–96h obtained for diploids in the present study was similar to the value of 0.52 mg O_2 L^{-1} that was observed by Braun and colleagues (2006) for jundia diploids.

The greater sensitivity of jundia triploids under hypoxia conditions can be associated with the decreased numbers of corporal cells observed in triploid fish. According to Benfey (1999), the decreased concentration of hemoglobin

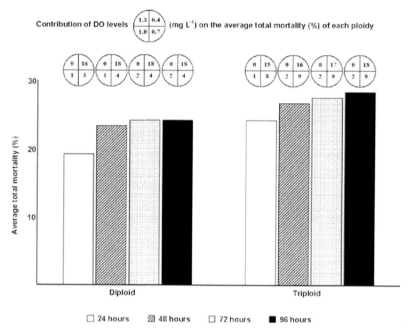

FIGURE 2 Total mortality average (%) for diploids and triploids of jundia *Rhamdia quelen* throughout the exposure period of the different DO concentrations tested. Circular graphs above each bar represent the cumulative mortality average (clockwise) for each ploidy during the experimental period.

TABLE 2 Mean LC50 Values for Diploids and Triploids of Jundia *Rhamdia Quelen* During 24-hour Intervals of Exposure to Different DO Concentrations, Including the Respective Confidence Interval (95%)

	Jundia Diploid			Jundia Triploid		
		Confidence intervals (95%)			Confidence intervals (95%)	
Hours	LC50 (mg L^{-1} DO)	Minimum (mg L^{-1} DO)	Maximum (mg L^{-1} DO)	LC50 (mg L^{-1} DO)	Minimum (mg L^{-1} DO)	Maximum (mg L^{-1} DO)
24	0.485	0.299	0.594	0.516	0.265	0.655
48	0.531	0.386	0.644	0.540	0.275	0.696
72	0.535	0.366	0.658	0.554	0.326	0.707
96	0.535	0.366	0.658	0.567	0.366	0.717

Values generated by the "Probit Analysis Program."

observed in triploids justifies the change in oxygen transport capacity of these organisms.

Comparing fish performance for both ploidys under different oxygen concentrations, triploids have presented reduced capacity for oxygen transportation in *Salmo salar* (Graham, Fletcher, &. Benfey 1985), lower oxygen consumption in *Salvelinus fontinalis* (Stillwell & Benfey 1996), and minor aerobic capacity in triploids of *Oncorhynchus mykiss* (Virtanen, Forsman, & Sundby 1990). This physiological condition results in the greatest mortality of *O. mykiss* triploids, which was observed under conditions of increasing water temperature and decreasing DO availability (Ojolick et al. 1995).

Different results have been registered for other fish species, such as the similar oxygen consumption by *Oncorhynchus masou macrostomus* (Nakamura et al. 1989) and *Plecoglossus altivelis*; however, triploids of *P. altivelis* have shown a tendency for the highest oxygen consumption rates (Aliah et al. 1990). This study demonstrated that triploids of jundia *Rhamdia quelen* juveniles had a greater sensitivity to hypoxia when compared with diploids.

REFERENCES

Aliah R.S., K. Yamaoka, Y. Inada, and N. Taniguchi. 1990. Effects of triploidy on tissue structure of some organs of ayu. *Nippon Suisan Gakkaishi* 56:569–575.

Aliah R.S., K. Yamaoka, Y. Inada, and N. Taniguchi. 1991. Effects of triploidy on hematological characteristics and oxygen consumption in ayu. *Nippon Suisan Gakkaishi* 57:833–836.

Barcellos, L.J.G., G.F. Wassermann, A.P. Scout, V.M. Whoel, R.M. Quevedo, I. Ittzés, M.H. Krieger, and F. Lulhier. 2001. Steroid profiles in cultured female jundiá, the Siluridae *Rhamdia quelen* (Quoy and Gaimard, Pisces Teleostei), during the first reproductive cycle. *Gen. Comp. Endocrinol.* 121:325–332.

Benfey, T.J. 1999. The physiology and behavior of triploid fishes. *Rev. in Fish. Sci.* 7(1): 39–67.

Bergheim, A., M. Gausen, A. Naess, P.M. Hoelland, P. Krogedal, and V. Crampton. 2006. A newly developed oxygen injection system for cage farms. *Aquacult. Eng.* 34:40–46.

Boyd, C.E. 1990. *Water quality in ponds for aquaculture*. Auburn University, AL: Birmingham Publishing Co.

Boyd, C.E., and B.J. Watten. 1989. Aeration systems in aquaculture. *Rev. Aquat. Sci.* 1:425–472.

Braun, N., R.L. Lima, B. Moraes, V.L.P. Vieira, and B. Baldisserotto. 2006. Survival, growth and biochemical parameters of silver catfish, *Rhamdia quelen*, juveniles exposed to different dissolved oxygen levels. *Aquacult. Res.* 37:1524–1531.

Budiño, B., R.M. Cal, M.C. Piazzon, and J. Lamas. 2006. The activity of several components of the innate immune system in diploid and triploid turbot. *Comp. Biochem. Physiol. A: Mol. Integr. Physiol.* 145:108–113.

Carneiro, P.C.F., F. Bendhack, J.D. Mikos, M. Schorer, and P.R.C. Oliveira Filho. 2002. Jundiá: um grande peixe para a região sul. *Panorama da Aqüicultura* 12: 41–46.

Chabot, D., and J. Dutil. 1999. Reduced growth of Atlantic cod in non-lethal hypoxic conditions. *J. Fish Biol.* 55:472–491.

Cooper, R.U., L.M. Clough, M.A. Farwell, and T.L. West. 2002. Hypoxia-induced metabolic and antioxidant enzymatic activities in the estuarine fish *Leiostomus xanthurus*. *J. Exp. Mar. Biol. Ecol.* 279:1–20.

Environmental Protection Agency (EPA). 1984. *Ambient water quality criteria for ammonia–1984*. Springfield, VA: National Technical Information Service.

Fast, A.W., T. Pewnim, R. Keawtabtim, R Saijit, F.T. Te, and R. Vejaratpimol. 1995. Comparative growth of diploid and triploid Asian catfish *Clarias macrocephalus* in Thailand. *J. World Aquacult. Soc.* 26:390–395.

Fracalossi, D.M., E. Zaniboni-Filho, and S. Meurer. 2002. No rastro das espécies nativas. *Panorama da Aqüicultura* 12:43–49.

Fracalossi, D.M., G. Meyer, F.M. Santamaria, M. Weingartner, and E. Zaniboni-Filho. 2004. Desempenho do jundiá, *Rhamdia quelen*, e do dourado, *Salminus brasiliensis*, em viveiros de terra na região sul do Brasil. *Acta Scientiarum, Animal Sciences, Maringá*, 26:345–352.

Golterman, H., R.S. Clymo, and M.A.M. Ohnstad. 1978. Methods for physical and chemical analysis of fresh water. Oxford: Blackwell.

Gomes, L.C., J.I. Golombieski, A.R. Chippari Gomes, and B. Baldisserotto. 2000. Biologia do jundiá *Rhamdia quelen* (Teleostei, Pimelodidae). *Ciência Rural* 30(1): 179–185.

Graham, M.S., G.L. Fletcher, and T.J. Benfey. 1985. Effect of triploidy on blood oxygen content of Atlantic salmon. *Aquaculture* 50: 133–139.

Greenburg, A. 1992. *Standard methods for the examination of water and wastewater*. 18 ed. Washington, DC: Am. Public Health Assoc.

Huergo, G.M., and E. Zaniboni-Filho. 2006. Triploidy induction in jundiá *Rhamdia quelen*, through hydrostatic pressure shock. *J. Appl. Aquacult.* 18(4): 45–57.

Karim, R., M. Sekine, and M. Ukita. 2002. Simulation of eutrophication and associated occurrence of hypoxic and anoxic condition in a coastal bay in Japan. *Mar. Poll. Bull.* 45:280–285.

Kavumpurath, S., and T.J. Pandian. 1992. Effects of induced triploidy on aggressive display in the fighting fish, *Betta splendens* Regan. *Aquacult. Fish. Manage.* 23:281–290.

Kerbya, J.H., J.M. Eversona, R.M. Harrellb, J.G. Geigerc, C.C. Starlingd, and H. Revelsd. 2002. Performance comparisons between diploid and triploid sunshine bass in fresh water ponds. *Aquaculture* 211:91–108.

Koroleff, F. 1975. Metodologia para Determinação da Amônia. In *Methods of seawater analysis*, ed. K Grasshoff, 117–181. New York: Verlag. Chemie. Weinheim

Madenjian, C.P., G.L. Rogers, and A.W. Fast. 1987. Predicting nighttime dissolved oxygen loss in prawn ponds of Hawaii: Part I. Evaluation of traditional methods. *Aquacult. Eng.* 6:191–208.

Maffezzolli, G., and A.P.O. Nuñer. 2006. Crescimento de alevinos de jundiá, *Rhamdia quelen* (Pisces, Pimelodidae), em diferentes concentrações de oxigênio dissolvido. *Acta Scientiarum Anim. Sci.* 28(1): 41–45.

Marchioro, M.I., and B. Baldisserotto. 1999. Sobrevivência de alevinos de jundiá (*Rhamdia quelen* Quoy e Gaimard 1824) à variação de salinidade da água. *Ciência Rural* 29:315–318.

Nakamura, S., N. Imai, S. Ohya, and H. Kobayashi. 1989. Oxygen uptake rate and hematological characteristics in triploid amago salmon, *Oncorhynchus masou macrostomus*. *Memoirs of the Faculty of Agriculture of Kinki University (Japan)*: 31–38.

Ojolick, E.J., R. Cusack, T.J. Benfey, and S.R. Kerr. 1995. Survival and growth of all-female diploid and triploid rainbow trout (*Oncorhynchus mykiss*) reared at chronic high temperature. *Aquaculture* 131:177–187.

Phillips, R.B., K.D. Zajicek, P.E. Ihssen, and O. Johnson. 1986. Application of silver staining to the identification of triploid fish cells. *Aquaculture* 54:313–319.

Silfvergrip, A.M.C. 1996. A systematic revision of the neotropical catfish genus *Rhamdia* (Teleostei, Pimelodidae). PhD thesis, Dept. of Vertebrate Zoology, Swedish Museum of Natural History, Stockholm, Sweden.

Small, S.A., and T.J. Benfey. 1987. Cell size in triploid salmon. *J. Exp. Zool.* 241: 339–342.

Stillwell, E.J., and T.J. Benfey. 1996. Hemoglobin level, metabolic rate, opercular abduction rate and swimming efficiency in female triploid brook trout (*Salvelinus fontinalis*). *Fish Physiol. Biochem.* 15:377–383.

Tiwary, B.K., R. Kirubagaran, and A.K. Ray. 2000. Gonodal development in triploid *Heteropneustes fossilis*. *J. Fish Biol.* 57:1343–1348.

Virtanen, E., L. Forsman, and A. Sundby. 1990. Triploidy decreases the aerobic swimming capacity of rainbow trout (*Salmo gairdneri*). *Comp. Biochem. Physiol.* 96A:117–121.

Wilhelm Filho, D., M.A. Torres, E. Zaniboni-Filho, and R.C. Pedrosa. 2005. Effect of different oxygen tensions on weight gain, feed conversion, and antioxidant status in piapara, *Leporinus elongatus* (Valenciennes, 1847). *Aquaculture* 244: 349–357.

Wolters, W.R., C.L. Chrisman, and G.S. Libey. 1982. Erythrocyte nuclear measurements of diploid and triploid channel catfish, *Ictalurus punctatus* (Rafinesque). *J. Fish Biol.* 20:253–258.

Wu, R.S.S. 2002. Hypoxia: From molecular responses to ecosystem responses. *Marine Poll. Bull.* 45:35–45.

Effect of Stock Density and Ploidy in Jundia, *Rhamdia quelen*, Larvae Performance

HIRLA FUKUSHIMA[1], JHON EDISON JIMENEZ[1],
MARCOS WEINGARTNER[2], and EVOY ZANIBONI-FILHO[2]

[1]*Aquaculture Graduate Program, Federal University of Santa Catarina, Florianopolis, Brazil*
[2]*Freshwater Fish Biology and Fish Culture Laboratory, Aquaculture Department, Federal University of Santa Catarina, Florianopolis, Brazil*

The performance (growth and survival) of diploid and triploid jundia, Rhamdia quelen, *was evaluated at six different stocking densities (10, 60, 110, 160, 210, 260 larvae/liter) during 31 days after rearing in an intensive larviculture system. Triploid fish exhibited a significantly higher survival rate than diploids at all stocking densities (27.1 ± 4.3% vs. 12.1 ± 3.3%; P < 0.01). Survival was not affected by stocking density (P > 0.05). Length gain was not affected by either ploidy or stocking density. Diploid fish gained more weight than triploids (P < 0.05), though this difference could result from lower fish densities in diploid treatments resulting from the higher mortality rate of diploid fish. This hypothesis is strengthened by the higher biomass present in triploid treatments (P < 0.01).*

INTRODUCTION

The jundia, *Rhamdia quelen* (Siluriformes: Heptapteridae) (Figure 1), is a freshwater catfish native to hydrographic basins from southeast México through central Argentina (Zaniboni Filho 2003). This species has great

---

The authors thank CNP for a grant to E. Zaniboni-Filho and Capes for a grant to H. Fukushima.

FIGURE 1 Jundia *Rhamdia quelen* (Siluriformes: Heptapteridae) (color figure available online).

economic potential for Brazilian aquaculture due to its promising zootechnical and organoleptic characteristics, benefiting both producers and the processing industry. This fish is marked by fast growth, good food conversion rate, stress resistance, and tolerance for low temperatures under cultivation, permitting continuous growth year-round (Fracalossi et al. 2004). Moreover, jundia is a popular food fish with flesh free of intramuscular bones.

The production of jundia is expanding in southern Brazil (Gomes et al. 2000). However, precocious sexual maturation has been identified as a potential problem (Fracalossi et al. 2004), as both sexes reach sexual maturation before reaching commercial weight (Baldisseroto & Neto 2004), which substantially decreases their growth rate.

The induction of triploidy is a chromosomal manipulation technique that has been used to induce sterility and thus avoid the problems of sexual maturation in many species of fish under commercial cultivation (Arai 2001) including carp, catfish, tilapia, and salmon (Belmont & Hoare 2003). Triploid fish have been reported to exhibit higher rates of survival, growth, food conversion, and disease resistance for triploids than diploids (Kerby et al. 2002). The sterility resulting from triploidy also prevents any genetic interactions between cultivated and wild fish (Cotter et al. 2000), avoiding any impact on the genetic and ecological structure of natural populations (Lutz 2001). Triploidy can be induced via thermal or pressure shocks to recently fertilized eggs, impeding metaphase II during meiosis (Dunham 2004). Recently, Huergo and Zaniboni Filho (2006) demonstrated the viability of producing and growing triploid *R. quelen* larvae.

With an established protocol for producing *R. quelen* triploids, methods for optimal triploid larviculture must be developed to enable the production of juveniles for large-scale cultivation. The present study evaluates the interaction between ploidy and larval stocking densities on growth and survival.

MATERIAL AND METHODS

Larvae were obtained from *R. quelen* broodstock descended from wild populations of the Uruguay River basin, and maintained by the Freshwater Fish

Biology and Fish Culture Laboratory of the Federal University of Santa Catarina in Florianopolis, Brazil. Eight female and two male *R. quelen* were treated with carp pituitary extract to induce spawning. Gametes were obtained during the 234th degree/hour (after 9 hours in 26°C) after the hormonal treatment and pooled. After 4 minutes of gamete fertilization, one group of eggs was subjected to a pressure shock, following the procedure used by Huergo and Zaniboni Filho (2006) to induce triploidy in this species. The eggs were placed in an 800-ml steel chamber, and pressure (5.000 psi) was applied for five minutes using a French press. The remaining eggs formed the diploid control group. The diploid and triploid eggs were maintained at 26°C in separate 56-L cylinder-conical incubators with a recirculation system guaranteeing constant water movement. Hatching occurred 36 hours after fertilization, and larvae were maintained in the incubators until mouth opening was observed and exogenous feeding began.

The experimental was arranged in a completely randomized 3 × 2 factorial design. After the incubation period, larvae of each ploidy were transferred to 36 experimental larviculture units. For each ploidy, six stocking densities ranging from 10 to 260 larvae per liter were tested in triplicate over 31 days. Simultaneously, 30 larvae of each ploidy were fixed in 4% bufered formalin for subsequent biometry. Larvae were fed *Artemia sp.* nauplii twice a day (at 08:00 and 16:00 hours) for the first 18 days of treatment, after which an *ad libitum* diet containing 56% crude protein and 3,700 Kcal/kg of available energy (Table 1).

The experimental units consisted of light-colored, 5-L rectangular tanks with constant aeration and biofiltration in a recirculating system. There is some research with South American freshwater fish that shows better survival and frequently better growth under larviculture with slightly salty water (Beux & Zaniboni-Filho 2007; Santos & Luz 2009). Each experimental unit was maintained at 26°C and continuously supplied with salt water (2.5 ppm NaCl) with a renewal rate of 200% per day.

The basic water quality parameters (pH, dissolved oxygen, temperature, and conductivity) of each experimental unit were monitored twice a day using a multiparameter probe (YSI63, Yellow Spring, OH, USA). Ammonia, nitrite, and hardness were measured weekly using colorimetric methods (Alpha Tecnoquímica, SC, Brazil). Mortality occurring in each treatment was quantified daily, and excrement and any leftover food was removed.

TABLE 1 Feeding Regimen of *Rhamdia quelen* Diploid and Triploid Larvae During the Experimental Period

Experimental period (days)	01–03	04–06	07–09	10–12	13–18	19–30
Feeding (nauplii/larvae)	50	100	150	200	300	400 + food[1]

[1]Diet containing 56% crude protein, 3700Kcal/kg available energy, 10% fat, 3.6% calcium, and 1.5% total phosphorus.

At the end of the experiment, samples from the fins of 30 fish from each triploid experimental unit (540 samples total) were collected and fixed for subsequent confirmation of triploidy. Triploidy was verified using the $AgNO_3$ nucleus coloration method (Phillips et al. 1986), confirming that all treated fish were triploids. Data was also collected to calculate the following parameters:

- Weight gain (g): WG = W $_{final\ average}$ − W $_{initial\ average}$;
- Length gain (mm): LG = L $_{final\ average}$ − L $_{initial\ average}$;
- Survival (%): S = 100 * N_f/N_i, where N_f = number of surviving fish at the end of the experimental period, and N_i = number of fish present at the beginning of the experimental period; and
- Daily mortality rate (%) = number of dead larvae on a given day/initial number of larvae * 100

An ANOVA (Sliced, SAS) was performed to identify interaction between stocking density and ploidy. These relationships were further explored via linear regression analyses and a post-hoc Tukey test ($\alpha < 0.05$) to verify differences between treatment averages.

RESULTS AND DISCUSSION

Larviculture is considered the critical phase in the production of juvenile native Brazilian fishes (Luz & Zaniboni Filho 2002), and stocking densities can affect the growth and survival of jundia larvae (Piaia & Baldisseroto 2000). The survival rate of jundia was greater for triploids than diploids across all tested stocking densities ($P < 0.01$), with survival averaging 120% higher in triploids (Figures 2 and 3). Overall survival of jundia, diploid, or triploid was not affected by stocking density ($P > 0.05$).

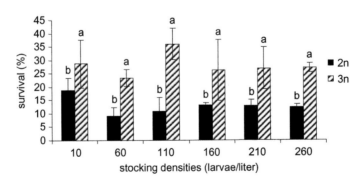

FIGURE 2 Survival rate (average ± standard deviation) of diploid and triploid *Rhamdia quelen* larvae reared at different stocking densities for 31 days.

Different letters indicate significant differences between survival rates ($P < 0.01$).

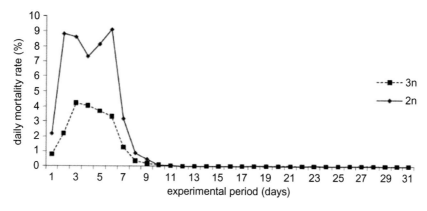

FIGURE 3 Average daily mortality rate of diploid and triploid *Rhamdia quelen* larvae during 31 days of intensive larviculture.

The stocking density recommended for diploid jundia larviculture is about 10 larvae/L (Baldisseroto & Neto 2004). Survival in our study was lower than expected. Silva (2004) achieved survival values between 80% and 95% during the first 21 days of jundia rearing. Survival rates are affected by many factors, including handling, brightness, disruptive human activities, and tank color (Behr et al. 1999). In this study, certain conditions differed from those used in typical jundia larviculture: tanks were of light color and low volume with an elevated rate of water circulation. These differences may have contributed to a reduced survival rate relative to the optimum, but as experimental conditions were identical among treatments, the results may be compared.

Fish maintained in high densities are generally exposed to a complex group of factors that also interact, including water quality, behavior alterations due to social interactions, and reduced food availability (Urbinati & Carneiro 2004). However, in the present study the available food was scaled with the biomass present in each treatment. Water quality was also kept nearly constant across all ploidies and stocking densities, as follows: temperature was maintained at $25 \pm 0.7°C$, dissolved oxygen concentration above 6 mg/L, pH at 7.6 ± 0.8, alkalinity at 103.0 ± 6.9 mg/L $CaCO_3$, ammonia below 0.5 mg/L, and nitrate below 0.01 mg/L.

As noted by Silva (2004), jundia larvae are cannibalistic, beginning with the initiation of exogenous feeding and increasing over the following six days. Similar behavior was observed in larvae of both ploidies during this study, extending through the seventh day of exogenous feeding and being more accented in diploids treatments. One effect of triploidy in fishes is the reduction of aggression (Carter et al. 1994; Garner et al. 2008; O'Keefe & Benfey 1997; Wagner et al. 2006), which suggests that the higher survival rates observed for triploid *R. quelen* may result from a reduction of the species' characteristic cannibalism during the initial phase of larvae development. Reduced aggression also prevents the formation of social hierarchies, decreasing socially induced physiological effects on submissive fish. These observed differences in behavior relate to a reduction in the number of brain and sensory cells present in triploids; in addition to reduced aggression, this results in lower sensitivity to light and sound (Aliah et al. 1990), decreasing the stress triploids experience during handling.

A higher weight gain was observed for diploids than for triploids ($P < 0.05$), while length gain was not affected by ploidy or stocking density ($P < 0.05$) (Table 2). Nevertheless, because of the higher survival rate among triploids, triploids attained a higher total biomass than diploids except at the lowest stocking density (10 larvae/L) ($P < 0.01$; Table 2). A positive correlation between biomass and stocking density was observed in both ploidies, with the increase of biomass in the triploids varying between 19% and 150% above diploids (Figure 4).

Several studies indicate that triploid fish rarely grow faster than diploids during the juvenile phase and that positive growth and food conversion benefits of triploidy occur at sexual maturation, when triploids present the highest (Wolters, Chrisman, & Libey 1982; Taniguchi et al. 1986; Dunham 1990). On the other hand, in some species triploids do present superior performance beginning in the initial phase of development (Woters et al. 1982; Taniguchi et al. 1986), presumably due to the advantage conferred by increased heterozygosity, as triploids have two sets of maternal chromosomes and one paternal set. This composition increases heterozygosity by 30%–60% compared to the homologous diploids (Dunham 2004), improving phenotypic performance according to the phenomenon of "hybrid vigor" (Beaumont & Hoare 2003). In other siluriform fishes, diploid-triploid growth

TABLE 2 Performance Results (Average ± Standard Deviation) of Diploid (2n) and Triploid (3n) *Rhamdia quelen* Larvae Under Different Stocking Densities (ED)

	Variables					
	Final biomass (g)[1]		Weight gain (g)[1]		Length gain (mm)[1]	
ED (larvae/L)	2n	3n	2n	3n	2n	3n
10	3.41 ± 0,87A	4.06 ± 1.43A	0.38 ± 0.02A	0.29 ± 0.03B	34.51 ± 0.90A	33.43 ± 0.42A
60	6.50 ± 2,6A	14.83 ± 6.8B	0.30 ± 0.02A	0.29 ± 0.03A	31.21 ± 0.82A	33.44 ± 1.28A
110	13.44 ± 9,8A	33.48 ± 4.73B	0.26 ± 0.02A	0.21 ± 0.02B	30.49 ± 1.3A	30.40 ± 0.50A
160	29.41 ± 2,74A	39.19 ± 8.7B	0.28 ± 0.05A	0.22 ± 0.03B	31.33 ± 1.95A	32.58 ± 3.37A
210	38.88 ± 4,41A	53.07 ± 8.06B	0.29 ± 0.02A	0.23 ± 0.07B	33.00 ± 0.12A	35.14 ± 0.93A
260	43.39 ± 3.01A	71.63 ± 5.40B	0.26 ± 0.02A	0.20 ± 0.02B	32.70 ± 0.70A	30.16 ± 0.80A
Mean[2]	22.5 ± 17.06A	36.04 ± 24.68B	0.30 ± 0.04A	0.24 ± 0.04B	32.21 ± 1.48A	32.53 ± 1.93A

[1]Different letters (A vs. B) indicate significant differences ($P < 0.05$) between ploidies reared at the same stocking density.
[2]Mean gives the average value for each ploidy across all tested stocking densities. Different capital letters indicate significant differences ($P < 0.05$) between the averages for diploids and triploids.

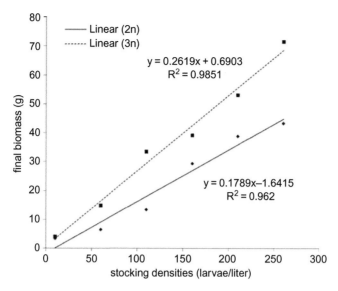

FIGURE 4 Relation between final biomass of diploids (2n) and triploids (3n) of jundia *Rhamdia quelen* and the different stocking densities during 31 days of intensive larviculture.

comparisons have yielded conflicting results, with some studies finding a triploid growth advantage (Wolters, Chrisman, & Libey 1982; Chrisman, Wolters, & Libey 1983; Krasznai & Marian 1986; Fast et al. 1995; Qin, Fast, & Ako 1998; Olufeagba, Aluko, & Omotosho 2000). Others have observed similar growth for both ploidies (Henken, Brunink, & Richter 1987; Richter et al. 1987; Lilystrom et al. 1999), or even inferior growth of triploids (Na-Nakorn & Legrand 1992). According to O'Keefe and Benfey (1999), this

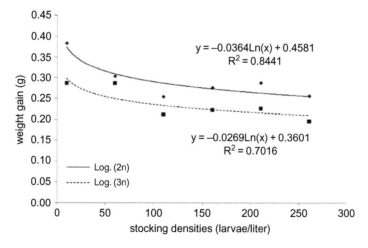

FIGURE 5 Relation between the weight gain of diploid and triploid *Rhamdia quelen* larvae reared at different stocking densities for 31 days.

divergence in growth results is due to the different effects of triploidy as it manifests in different species, rearing conditions, and the methods used to induce triploidy.

Our ability to compare growth rates within stocking densities was damaged by the difference in the survival rate observed between the ploidies, as different mortality rates affected the densities experienced by the surviving fish. The negative correlation between stocking density and growth is widely known (Zaniboni Filho 2000; Inoue et al. 2003) and is supported herein (Figure 5).

REFERENCES

Aliah R.S., K. Yamaoka, Y. Inada, and N. Taniguchi. 1990. Effects of triploidy on tissue structure of some organs of ayu. *Nippon Suisan Gakkaishi* 56: 569–575.

Arai, K. 2001. Genetic improvement of aquaculture finfish species by cromossome manipulation techniques in Japan. *Aquacult.* 197:205–228.

Baldisserotto, B., and J.R. Neto. 2004. Biologia do jundiá. In *culture of Jundiá,* edited by B. Baldisserotto, and J.R. Neto, 67–72. Santa Maria, Brazil: UFSM.

Beaumont, A., and K. Hoare. 2003. *Biotechnology and genetics in fisheries and aquaculture*. Bangor, UK: Wiley/Blackwell.

Behr, E.R., J. Radunz Neto, A.P. Tronco, and A.P. Fontana. 1999. Influence of different light intensity on the performance of larvae of jundiá (*Rhamdia quelen*) (Quoy and Gaimard, 1824) (Pisces: Pimelodidade). *Acta Scientiarum* 21:325–330.

Beux & Zaniboni-Filho. 2007. Survival and the growth of pintado (*Pseudoplatystoma corruscans*) post-larvae on different salinities. *Braz. Arch. Biol. Technol.* 50(5): 821–829.

Carter, C.G., I.D. Mccarthy, D.F. Houlihan, R. Johnstone, M.W. Walsingham, and A.I. Mitchell. 1994. Food consumption, feeding behavior, and growth of triploid and diploid Atlantic salmon, *Salmo salar* L., parr. *Can. J. Zool.* 72:609–617.

Chrisman, C.L., W.R. Wolters, and G.S. Libey. 1983. Triploidy in channel catfish. *J. World Maricult. Soc.* 14:279–293.

Cotter, D., V. O'Donovan, N. O'Maoiléidigh, G. Rogan, N. Roche, and N.P. Wilkins. 2000. An evaluation of use of triploid atlantic salmon (*salmo salar* L.) in minimizing the impact of escaped farmed salmon on wild populations. *Aquacult.* 186:61–75.

Dunham, R.A. 1990. Genetic engineering in aquaculture. *AgBiotech News Inf.* 2:401–406.

Dunham, R.A. 2004. *Aquaculture and fisheries biotechnology: Genetics approaches.* Tallahasse. FL: CABI.

Fast, A.W., T. Pewnim, R. Keawtabtim, R. Saijit, F.T. Te, and R. Vejaratpimol, R. 1995. Comparative growth of diploid and triploid asian catfish *Clarias macrocephalus* in Thailand. *J. World Aquacult. Soc.* 26:390–395.

Fracalossi, D.M., G. Meyer, F.M. Santamaria, M. Weingartner, and E.Zaniboni Filho. 2004. Performance of jundiá, Rhamdia quelen, and of dourado, *Salminus brasiliensis*, cultivated in ponds in the South region of Brazil. *Acta Scientiarum* 26:345–352.

Garner, S.R., B.N. Madison, N.J. Bernier, and B.D. Neff. 2008. Juvenile growth and aggression in diploid and triploid Chinook salmon *Oncorhynchus tshawytscha* (Walbaum). *J. Fish Biol.* 73:169–185.

Gomes, L.C., J.L. Golombieski, A.R.C. Gomes, and B. Baldisserotto. 2000. Biology of jundiá *Rhamdia quelen* (Teleostei, Pimelodidae). *Ciência Rural* 30:179–185.

Henken, A.M., A.M. Brunink, and C.J.J. Richter. 1987. Differences in growth rate and feed utilization between diploid and triploid African catfish, *Clarias gariepinus* (Burchell, 1822). *Aquacult.* 63:233–242.

Huergo, G.P.C.M., and E. Zaniboni Filho. 2006. Triploidy induction in Jundiá, *Rhamdia Quelen* through hydrostatic pressure. *J. Appl. Aquacult.* 18:45–57.

Inoue, L.A.K.O., J.A. Senhorini, and E.Zaniboni Filho. 2003. Growth of pacu juveniles in nightly aerated system. *Acta Scientiarum* 25:45–48.

Kerby, H., M. Everson, M. Harrell, R. Geiger, C. Starling, and H. Revels. 2002. Performance comparisons between diploid and triploid sunshine bass in fresh water ponds. *Aquacult.* 211:91–108.

Krasznai, Z., and T. Marián. 1986. Shock-induced triploidy and its effect on growth and gonad development of European catfish, *Silurus glanis* L. *J. Fish Biol.* 29:519–527.

Lilystrom, C.F., W.R. Woltwrs, D. Bury, M. Rezk, and R.A. Dunham. 1999. Growth, carcass traits and oxygen tolerance of diploid and triploid catfish hybrids. *North Am. J. Aquacult.* 61:293–303.

Lutz, C.G. 2001. Pratical genetics for aquaculture. Malden, MA: Fishing News Books.

Luz, R.K., and E.Zaniboni Filho. 2002. Larviculture of mandi-amarelo *Pimelodus maculatus* LACÉPÈDE, 1803 (Siluriformes: Pimelodidade) in different densities in the first days of life. *Revista Brasileira de Zootecnia* 31:560–565.

Na-Nakorn, U., and E. Legrand. 1992. Induction of triploidy in *Puntius gonionotus* (Bleeker) by cold shok. *Kasetsart Univ. Fish. Res. Bull.* 18:10.

O'Keefe, R.A., and T.J. Benfey. 1999. Comparative growth and food consumption of diploid and triploid brook trout (*Salvelinus fontinalis*) monitored by radiography. *Aquacult.* 175:111–120.

Olufeagba, S.O., P.O. Aluko, and J.S. Omotosho. 2000. Effect of triploidy on fertility of African catfish *heterobranchus longifilis* (Family: Clariidae). *J. Fish. Technol.* 2:43–50.

Phillips, R.B., K.D. Zajicek, P.E. Ihssen, and O. Johnson. 1986. Application of silver staining to the identification of triploid fish cells. *Aquacult.* 54:313–319.

Piaia, R., and B. Baldisserotto. 2000. Stock density and growth in juveniles of jundiá *Rhamdia quelen* (Quoy e Gaimard 1824). *Ciência Rural* 30:509–513.

Qin, J.G., A.W. Fast, and H. Ako. 1998. Growout performance of diploid and triploid Chinese catfish *Clarias fuscus*. *Aquacult.* 166:247–258.

Richter, C.J.J., A.M. Henken, E.H. Eding, J.H. Van Doesum, and P. Boer. 1987. Induction of triploidy by cold-shoking eggs and performance of triploids in African catfish *Clarias gariepinus* (Burchell, 1822). In *Selection, hybridization, and genetic engineering in aquaculture*, edited by K. Tiews, 225–237. Berlin, German: Heeneman Verlagsgesellschaft.

Santos & Luz. 2009. Effect of salinity and prey concentrations on *Pseudoplatystoma corruscans, Prochilodus costatus,* and *Lophiosilurus alexandri* larviculture. *Aquacult.* 287:324–328.

Silva, L.V.F. 2004. Reprodução. In *Culture of jundiá*, edited by B. Baldisserotto and J. Radunz Neto, 95-106. Santa Maria, Brazil: UFSM.

Taniguchi, N., A. Kijina, T. Tamura, K. Takegami, and I. Yamazaki. 1986. Color growth and maturation in ploidy manipulated fancy carp. *Aquacult.* 57:321–328.

Urbinati, E.C., and P.C.F. Carneiro. 2004. Práticas de manejo e estresse dos peixes em piscicultura. In *special topics in intensive freshwater fish culture*, edited by J.E.P. Cyrino, E.C. Urbinati, D.M. Fracalossi, and N. Castagnolli, 172–193. São Paulo, Brazil: TechArt.

Wagner, E.J., R.E. Arndt, M.D. Routledge, D. Latremouille, and R.F. Mellenthin. 2006. Comparison of hatchery performance, agonistic behavior, and poststocking survival between diploid and triploid rainbow trout of three different Utah strains. *North Am. J. Aquacult.* 68:63–73.

Wolters, W. R., C.L. Chrisman, and G.S. Libey. 1982. Erythrocyte nuclear measurements of diploid and triploid channel catfish, *Ictalurus punctatus* (Rafinesque). *J. Fish Biol.* 20:253–258.

Zaniboni Filho, E. 2000. Freshwater fish larviculture. *Informe Agropecuário* 21: 69–77.

Zaniboni Filho, E. 2003. Piscicultura das espécies nativas de água doce. In *Aqüicultura: Brazilian experiences*, edited by C.R. Poli, A.T.B. Poli, E.R. Andreatta, and E. Beltrame, 337–368. Santa Catarina, Brazil: Multitarefa.

Integrated Fish Farming for Nutritional Security in Eastern Himalayas, India

B. P. BHATT[1], K. M. BUJARBARUAH[2], K. VINOD[3], and M. KARUNAKARAN[4]

[1] Indian Council of Agriculture Research (ICAR) Research Complex for Eastern Region, ICAR Parisar, Bihar, India
[2] Indian Council of Agricultural Research (ICAR), Government of India, New Delhi, India
[3] ICAR Research Complex for NEH Region, Umiam, Meghalaya, India
[4] ICAR Research Complex for NEH Region, Nagaland Centre, Medziphema, Nagaland, India

An experiment was conducted in the North Eastern Himalayan region of India to evaluate the productivity of different animal-fish integrations viz. duck-cum-fish, poultry-cum-fish, goat-cum-fish, pig-cum-fish, and cattle-cum-fish. Among various integrations, cattle-cum-fish had the maximum fish yield (2,686.0 kg/ha), followed by fish-cum-duck (2,173.0 kg/ha), fish-cum-pig (2,046.0 kg/ha), fish-cum-poultry (2,006.0 kg/ha), and fish-cum-goat (1,867.0 kg/ha) compared to 757.0 kg/ha in unfed/unfertilized control ponds. Among fish species, highest growth rate was recorded for silver carp (Hypopthalmichthys molitrix) in all the integrations (201.0 to 591.0 g/fish), followed by mrigal (Cirrhinus mrigala, 160.0 to 370.0 g/fish). The implications of the results for development of suitable integrated agro-aquaculture models in humid tropics are discussed.

The authors express their gratitude to the Indian Council of Agricultural Research, Government of India, New Delhi, for financial assistance in the form of a research grant.
Views expressed in this article are of the authors and not necessarily of the Institute.

INTRODUCTION

Integrated farming of fish and livestock is an old practice. The aim of integrated farming is the recycling of animal wastes (feces, urine, and spoiled feeds) to serve as fertilizers and sometimes as food for fishes grown in ponds, enclosures, or cages (Edwards et al. 1986). The basic principle involved in integrated farming is the harnessing of complementarities of crops, livestock, and fishes, including recycling of farm renewable resources and natural resource conservation (Edwards et al. 1986; Bhatt, Bujarbaruah, & Vinod 2006).

The North Eastern Himalayan (NEH) region of India is in the humid tropics. It covers seven states of India. Most of the populations are non-vegetarians with pork, chicken, mutton, and beef as the major sources of animal protein, though fish consumption varies among the states from a low of 0.09 kg/capita/yr in Sikkim to a high of 9.22 kg/capita/yr in Tripura. Fish availability in Assam and Manipur was estimated to be 6.01 and 6.72 kg/capita/yr, respectively. In rest of the states, fish availability ranged from 2.29 to 3.21 kg/capita/yr.

The region receives very high precipitation (2,000 to 3,500 mm annual rainfall), which offers unique opportunity for fish farming. On average, the NEH region has 19,150 km of rivers; 23,792 ha reservoirs; 143,740 ha beels, lakes, and swamps; 40,809 ha of ponds/mini barrages; and 2,780 hectares of paddy-cum-fish culture areas (Mahapatra, Vinod, & Mandal 2004). It produces 223,000 mt of fishes against a projected requirement of 428,700 mt with a deficit of 48.0% (Bhatt, Bujarbaruah, & Vinod 2006). Due to this fact, fishes are being imported from outside the region, mainly from Andhra Pradesh. Keeping this fact in view, an attempt has been made in the present investigation to develop different integrated fish farming systems so as to find out the best integrations for increasing fish production in the region.

MATERIALS AND METHODS

The experiment was conducted over the period from 2001 to 2007 at an experimental farm of the Indian Council of Agricultural Research (ICAR), Meghalaya, India, situated between 25°39'–25°41'N latitude, 91°54'–91°63'E longitude and 940m asl altitude. The climate of the area was humid subtropical with an annual rainfall of 2,320 mm. On average, 90% of the total rainfall was received from April to October. The mean minimum and maximum temperature was recorded at 4°C during January and 28°C during July, respectively. Soils of the experimental plot were sandy loam, characterized by P-deficient acidic Alfisol with soil pH of 5.2 to 5.5 (Majumdar et al. 2004).

A total of six ponds covering an area of 8,100 m² was built to rear fishes integrated with ducks (Indian Runner), poultry (White Leghorn), goat

(Black Bengal), pig (Large Black), and cattle (Holstein Friesian). Slaked lime, $Ca(OH)_2$, was added to each pond at 500 kg/ha to maintain the pH value of 7.0 to 7.5. In one of the ponds, fishes were reared without integration of livestock/birds to serve as control. In the control pond, no fertilizer was added except liming so as to maintain the pH. The slopes of each pond were stabilized with grasses (Congo, guinea, and Napier), rice bean, and hedgerow species. On the pond dykes, fruit trees were planted at a distance of 5 m × 5 m from plant to plant (single row).

A polyculture of fishes was stocked in each pond at 9,000 fingerlings/ha with species composition of *Catla catla* (catla 20%), *Labeo rohita* (rohu, 10%), *Cirrhinus mrigala* (mrigal 20%), *Hypophthalmichthys molitrix* (silver carp 20%), *Ctenopharyngodon idella* (grass carp 20%), and *Labeo gonius* (gonius, 10%) in April and harvested during December every year. Average weight and length of fish fingerlings at stocking (Table 1) varied significantly among species ($P < 0.05$).

To integrate livestock with fishery, duck shed (3.6 m × 3.6 m × 1.8 m), poultry shed (3.0 m × 2.5 m × 1.8 m), goat shed (3.6 m × 3.6 m × 1.8 m), pig shed (2.4 m × 1.8 m × 1.5 m), and cattle shed (11.0 m × 10.0 × 4.2 m) were made with locally available bamboo material, roofed with corrugated galvanized iron (CGI) sheets and cement concrete floor on the dyke of the respective pond. Periodic growth of fish and livestock was recorded.

Stocking density of livestock for integrated fish farming varies with the environmental conditions of the locality, altitude, water quality, and fish composition *viz.* number of pigs per ha of pond area has been reported from 40 to 300; however, the number of piglets recommended is generally 100 per ha or 1 piglet per 100 m^2 of pond (New 1991). Similarly, for Asian countries, chicken and duck density has been reported to be 1,000–6,000 per ha (Pillay 1980). However, stocking density of livestock/birds had not been standardized for mid altitudes of the humid tropics in India where the study was conducted. Thus, in the present study, stocking density of goat, duck, poultry, pig, and cattle was standardized at 55, 500, 450, 34, and 5 animals per ha of fishpond, respectively (Bhatt, Bujarbaruah, & Vinod 2006).

TABLE 1 Growth of Fish Fingerlings During Stocking in Various Integrated Fish Farming Models, NEH Region, India

Sl. No.	Fish species	Common name	Length (cm)	Weight (g/fish)
1	*Catla catla*	Catla	5.52[abc]	14.27[b]
2	*Labeo rohita*	Rohu	5.91[a]	14.29[b]
3	*Cirrhinus mrigala*	Mrigal	5.07[bc]	9.90[d]
4	*Labeo gonius*	Gonius	4.80[cd]	11.24[c]
5	*Hypophthalmichthys molitrix*	Silver carp	6.36[a]	17.26[a]
6	*Ctenopharyngodon idella*	Grass carp	3.95[d]	9.21[d]
Duncan's multiple critical range ($P < 0.05$)			1.133	1.303

Means followed by the same letter are not significantly different ($P < 0.05$).

Dung/droppings of each category of animals were collected, weighed, and added to the respective ponds through drainage channel. Fecal material of livestock/birds was collected by pan washing method once a week at 8.00 A.M., and a representative sample was used for estimation of ash, dry matter, phosphorus, potassium, nitrogen, and calcium. Analysis of fecal material was done following standard procedures (Anderson & Ingram 1989), and total quantity added to the ponds was multiplied by the values obtained to record the total input of different nutrients.

Water samples of each pond were analyzed every month for water temperature, pH, alkalinity, hardness, dissolved oxygen, free carbon dioxide, and chloride using standard methods (APHA 1989). Plankton samples (phytoplankton and zooplankton) were collected (during daytime, i.e., 12 noon) by filtering 50 L of water from each pond using a plankton net having 173 meshes per linear inch (APHA 1989).

At the end of the study, fish flesh was subjected to proximate analysis by species regardless of integration. Observations were made for fish productivity under each category of animal husbandry and fish productivity over a period of seven years; hence, years were considered as replicates for each of the integrations. The data was statistically analyzed using one-way ANOVA and Duncan's multiple range test.

RESULTS AND DISCUSSION

The analyses were subjected to one-way ANOVA for both the variables, i.e., fish length and weight. Significant ($P < 0.05$) variations were observed for fish length and weight between species. Duncan's multiple critical range was obtained at 1.133 and 1.303, respectively, to fish length and weight. The results indicated highest fish length and fish weight for Silver carp, followed by rohu and catla (Table 1).

Annual input of nutrients to fish ponds by fecal material of different categories of animals has been worked out. Among various sources of nutrients, poultry manure exhibited highest nitrogen (2.12%) and phosphorus (1.40%), followed by dung droppings (1.85% nitrogen and 1.46% phosphorus, respectively). Potassium content was also highest in poultry manure (1.16%), followed by goat manure. Goat manure had highest calcium content (3.20%), followed by cow dung (2.46%). Lowest nitrogen (0.50%), potassium (0.28%), and phosphorus (0.35%) were, however, recorded in pig manure. On average, dry matter content was highest (50.16%) in cow dung and lowest (22.10%) in duck droppings. While cow dung contributed maximum phosphorus, nitrogen, and calcium into the pond, it was poultry manure that contributed maximum potassium.

Based on total dung/droppings added to the ponds, nitrogen (22.94 kg/yr) and phosphorus (15.15 kg/yr) contributions were highest in

TABLE 2 Fertilization of Fishponds (Kg/yr)* by Fecal Material of Different Categories of Livestock and Birds

Livestock component	Total dry weight of dung/droppings	Nitrogen	Phosphorus	Potassium	Calcium	Magnesium
Duck	586.66	10.85	8.56	5.75	11.67	13.02
Poultry	1,081.92	22.94	15.15	12.55	25.10	26.51
Goat	325.51	3.78	1.30	3.55	10.42	11.88
Pig	489.60	2.45	1.71	1.37	10.53	13.95
Cow	1,022.70	8.28	5.32	7.67	25.16	28.33

*Dry matter production was, respectively, 29.1, 92.0, 150.7, 1020.0, and 4870.0 g/day to duck, poultry, goat, pig, and cow dung. Poultry and cow dung was added to the ponds only for seven months in a year. Similarly, pig dung was added into the pond for eight months in a year.

poultry manure, followed by duck droppings. Potassium content was also highest in poultry manure (12.55 kg/yr), followed by cow dung (7.67 kg/yr). Calcium and magnesium was highest in cow dung, followed by poultry manure. Pig manure added the lowest amount of nitrogen, potassium, and calcium. Goat manure contributed the lowest amounts of phosphorus and magnesium (Table 2).

The pH in different integrations ranged from 7.2 to 7.6 compared to 7.26 in the control. Daily average high water temperature (recorded daily at 06:00, 12:00, and 18:00 hrs in all ponds) over the entire year was between 24.72 and 24.83°C in integrated ponds compared to 24.41°C in the control. Minimum temperature was between 22.01 and 22.50° C for all ponds, including the control.

Survival percentage for different fish species ranged from 80.0% to 95.0%. Overall, total fish productivity was 2.06-fold higher in integrations compared to the control. Of all the integrations, highest productivity was recorded for silver carp (three- to fourfold higher) compared to other fish species. Averaged for all years and all integrations, silver carp gained an average monthly body weight of 70.83 g. Fish productivity was recorded highest in fish-cum-cattle subsystem (2686.0 kg/ha), followed by fish-cum-duck (2173.0 kg/ha). Fish productivity was 2,010.0, 1,870.0, and 2,050.0 kg/ha, respectively, in fish-cum-poultry, fish-cum-goat, and fish-cum-pig integrations. Without integration (control), fish productivity averaged 757.0 kg/ha. Total fish production was 3.55-, 2.87-, 2.66-, 2.47-, and 2.71-fold higher when integrated with cattle, duck, pig, poultry, and goat, respectively, compared to control (Table 3).

These findings are in agreement with the earlier findings of Bujarbaruah and Bhatt (2006). Fish productivity was at par with the observations of earlier workers for mid altitudes of the NEH region, which face severe diurnal fluctuations in day and night temperatures (Bhatt, Bujarbaruah, & Vinod 2006; Mahapatra, Vinod, & Mandal 2003). For the Indian plain zones, however,

TABLE 3 Fish Productivity (Kg) Under Integrated Fish Farming

Fish species	Integrations					Control (Fishery without integration)
	Fish-cum-duck	Fish-cum-poultry	Fish-cum-goat	Fish-cum-pig	Fish-cum-cattle	
Catla	67.83	59.98	48.18	41.68	79.28	19.40
Sliver carp	86.30	83.20	73.90	75.40	82.02	24.19
Grass carp	56.10	32.90	31.13	44.60	69.30	12.10
Mrigal	67.50	30.18	36.00	48.61	53.91	18.41
Rohu	28.97	22.23	19.45	20.53	26.15	7.30
Gonius	19.29	12.25	15.39	14.64	11.61	5.62
Total yield (kg)	325.99	240.74	224.05	245.46	322.27	87.02
Area (m^2)	1,500.0	1,200.0	1,200.0	1,200.0	1,200.0	1,150.0
Productivity (kg/ha)	2,173.0	2,006.0	1,867.0	2,046.0	2,686.0	757.0

fish production in integrated farming systems was reported to be 5.0–7.5 mt/ha/yr (Samra et al. 2003).

Proximate analysis was conducted by species irrespective of integration and was not significantly different among species (Table 4). On average, chickens and ducks laid 95 and 75 eggs/bird/yr, respectively. Average egg production was, respectively, 5,320 and 3,410/yr, from chickens and ducks (Table 5).

At the time of slaughtering, the average live body weight was 25.2, 123.1, 2.89, and 2.25 kg/animal, for goats, pigs, ducks, and layer hens, respectively. Average daily weight gain was 4.01 gram/day or 0.004 kg/day–1.46 kg/yr for ducks; 340 gram/day or 0.340 kg/day– 124.10 kg/yr for pigs; 34.5 gram/day or 0.0345 kg/day–12.59 kg/yr for goats; and 3.125 gram/day or 0.0031 kg/day–1.1315 kg/yr for layer hens. Total meat production for the entire system was, respectively 195, 121, 82, and 35 kg from pigs, ducks, chickens, and goats. The slaughtering period was one year for fattening pig since its rearing thereafter was not economically viable (Bhatt & Bujarbaruah 2005; Kumaresan et al. 2007). However, in case of other category of animals, the slaughtering period was more than two years except for cows, which were reared as milch animals. So far as fish growth was concerned, silver carp exhibited maximum production in each integration followed by catla except in fish-cum-pig integration. Even in control pond, silver carp exhibited highest productivity.

Animal husbandry systems of tribal societies are organized on the basis of resource recycling within and between the forest and agriculture (Maikhuri 1992). If reared in an integrated fashion with judicious combination of fish-cum-livestock with other elements of agriculture, the overall productivity could increase as evidenced by the present investigation.

TABLE 4 Proximate Composition of Fish Species Averaged Over all Integrations

Fish species	Crude protein (g)	Fat (g)	Energy (K cal)	Ash (g)	Moisture (%)
Catla	18.98 (18.0–19.5)	2.16 (1.5–2.4)	5.66 (4.6–6.2)	1.55 (1.42–1.92)	74.01 (72.8–75.3)
Rohu	18.01 (16.0–18.8)	1.41 (1.1–1.9)	5.85 (5.7–6.1)	1.43 (1.1–2.0)	75.92 (72.1–77.7)
Mrigal	19.01 (18.3–19.4)	1.15 (0.7–1.4)	5.75 (5.1–6.1)	1.24 (0.9–1.6)	73.5 (70.5–76.9)
Grass carp	18.69 (18.3–19.1)	1.65 (1.5–2.1)	5.89 (5.5–6.2)	1.65 (1.1–2.0)	75.96 (75.3–77.2)
Silver carp	18.42 (18.1–19.0)	1.47 (1.3–1.6)	5.65 (5.4–5.9)	1.64 (1.2–2.1)	77.33 (76.3–79.11)
Gonius	18.83 (18.4–19.3)	1.56 (1.5–1.6)	6.16 (6.1–6.2)	1.80 (1.5–2.1)	72.07 (70.3–73.9)

Figures in parenthesis represent the range of maximum and minimum values.

TABLE 5 Productivity of Integrated Fish Farming

Sl. No.	Integrated systems	Area (m²)	Productivity per yr	
1	Fish-cum-duck	1900	Fish-	326 kg
			Duck egg-	341 nos.
			Duck meat-	121 kg*
2	Fish-cum-poultry	1700	Fish-	240 kg
			Egg-	5320 nos.
			Dressed chicken	82 kg*
3	Fish-cum-goat	1630	Fish-	224 kg
			Meat-	35 kg
4	Fish-cum-pig	1560	Fish-	245 kg
			Meat-	270 kg
5	Fish-cum-cattle	1960	Fish-	322 kg
			Milk-	3132 L
			FYM-	9060 kg
			Liquid manure-	16.3 m³
6	Fish without integration (control)	1550	Fish-	87 kg

*After two years.
Stocking density was 30, 55, 500, 450, and 05 per ha, respectively, to pig, goat, duck, chicken, and cattle. The dressing percentage was 75, 56, 58, and 70, respectively, to pig, goat, duck, and layer chicken.

REFERENCES

Anderson, J.M., and J.S.I. Ingram. 1989. *Tropical soil biology and fertility: A handbook of methods*. Wallingford, UK: CAB International.

APHA. 1989. *Standard methods for the examination of water and wastewater*. Washington, DC: American Public Health Association.

Bhatt, B.P., and K.M. Bujarbaruah. 2005. *Intensive integrated farming system: A sustainable approach of land use in eastern Himalayas, technical bulletin no 46*. Meghalaya, India: ICAR Research Complex for NEH Region, India.

Bhatt, B.P., K.M. Bujarbaruah and K. Vinod. 2006. *Integrated fish farming in eastern Himalayas, technical bulletin no. 47*. Meghalaya, India: ICAR Research Complex for NEH Region, India.

Bujarbaruah, K.M., and B.P. Bhatt. 2006. Complementarity of livestock and agroforestry for sustainable agriculture in North East. In *Agroforestry in northeast India: Opportunities and challenges*, edited by B.P. Bhatt, K.M. Bujarbaruah, 581–591. Meghalaya, India: ICAR Research Complex for NEH Region.

Edwards P., K. Kaewpaitoon, E.W. McCoy, and C. Chantachaeng. 1986. *Pilot small-scale crop/livestock/fish integrated farm, AIT research report no. 184*. Bangkok, Thailand: Asian Institute of Technology.

Kumaresan, A., K.M. Bujarbaruah, K.A. Pathak, B. Chhetri, S.K. Das, A. Das, and S.K. Ahmed. 2007. Performance of pigs reared under traditional tribal low input system and chemical composition of non-conventional tropical plants used as pig feed. *Livestock Sci*. 107:294–298.

Mahapatra, B.K., K. Vinod, and B.K. Mandal. 2003. Fisheries in northeast India and norms for sustainable development- an evaluation. In *Approaches for increasing agricultural productivity in hill and mountain ecosystem*, edited by B.P.

Bhatt, K.M. Bujarbaruah, Y.P. Sharma, and Patiram, 387–396. Meghalaya, India: ICAR Research Complex for NEH Region.

Mahapatra, B.K., K. Vinod, and B.K. Mandal. 2004. Fish biodiversity of north eastern India with a note on their sustainable utilization. *Environ. Ecol.* 22:56–63.

Maikhuri, R.K. 1992. Eco-energetic analysis of animal husbandry in traditional societies of India. *Energy* 17(10): 959–967.

Majumdar, B., K. Kumar, M.S. Venkatesh, Patiram and B.P. Bhatt. 2004. Effect of different agroforestry systems on soil properties in acid Alfisols of Meghalaya. *J. Hill Res.* 17(1): 1–5.

New, M.B. 1991. Turn of the millennium aquaculture, navigating troubled waters of riding the crest of the wave? *World Aquacult. Sci.* 22(3): 28–49.

Pillay, T.V.R. 1980. *Aquaculture: Principles and practices (Fishing New Books)*. New York: Wiley/Blackwell.

Samra, J.S., N. Sahoo, S.R. Chowdhury, R.K. Mohanty, S.K. Jena, and H.N. Verma. 2003. *Sustainable integrated farming system for waterlogged areas of eastern India*. Bhubaneshwar, India: Water Technology Centre for Eastern Region (ICAR).

Wastewater Aquaculture by the Mudialy Fisherman's Cooperative Society in Kolkata, West Bengal: An Example of Sustainable Development

ARCHANA SENGUPTA[1], TAPASI RANA[2], BISWAJIT DAS[1], and SHAMEE BHATTACHARJEE[3]

[1] *Vivekananda Institute of Medical Sciences, Kolkata, India*
[2] *Chittaranjan National Cancer Institute, Kolkata, India*
[3] *West Bengal State University, Department of Zoology, Barasat, India*

At the heart of one of the most densely populated mega-cities of India lies an eco-friendly and sustainable developmental business model. The city in question is Kolkata and the profitable but environmentally friendly venture that will be highlighted in this paper is that of wastewater-fed aquaculture practiced by a cooperative run by the local fishermen, namely Mudialy Fisherman's Cooperative Society (MFCS). Initially the marshy wetlands of Mudialy used to be flooded with polluted waste water from the city and the local industrial complex. Ultimately, the polluted water was released to the sacred River Ganges. This cooperative society has radically transformed these wetlands into an urban fishery and a water-front recreational ecosystem. The present paper focuses on the incredible task completed by the MFCS and also reports the results of water-quality analyses to ascertain that conditions in the water bodies of MFCS are ideal for a profitable aquaculture practice. All the water quality parameters studied (physicochemical water parameters, concentration of different heavy metals, and zooplankton community structure) were well within the normal limits from the point of view of pisciculture. This proves the efficiency of the wastewater purification system adopted by the cooperative society. The major advantage of wastewater-fed aquaculture over conventional wastewater treatment is the large diversity of marketable products and therefore broad possibilities of income generation. Thanks to the poor fishermen of MFCS, the water drained from these wetlands into the River Ganges is no longer polluted.

INTRODUCTION

The rapid growth of the human population, as well as the technological and industrial boom, has brought enormous problems and degradation of the environment. The process of urbanization in India since the beginning of last century reveals a steady increase in the size of its urban population. Pivotal to this urbanization phenomenon are the associated problems of providing municipal services and water resources. Ample supplies of clean unused water can no longer be taken for granted due to population growth, increasing urbanization, and industrial water demands. Pollution of freshwater streams and ground water by industrial discharges results in depletion of existing water sources. Hence, it has increasingly becoming obvious that reuse of wastewater is a viable solution in many instances. Water reclamation and reuse provides a unique and viable opportunity to augment traditional water supplies. As a multidisciplined and important element of water resources development and management, water reuse can help to close the loop between water supply and wastewater disposal.

Intense efforts are being made to treat domestic sewage to make the effluents suitable for discharge into the natural waters. The traditional practice of recycling sewage through agriculture, horticulture, and aquaculture—basic biological processes—has been in vogue in several countries. Taking ideas from these practices and deriving information from databases in different disciplines on wastewater management, aquaculture is being proposed and standardized as a tool for treatment of domestic sewage (Central Institute of Freshwater Aquaculture 1998). However, use of reclaimed water as an alternative poses concerns regarding its suitability for sustained development because of various issues related to wastewater usage and application. For instance, enrichment of nutrients in water bodies would normally increase fish production, but if it exceeds a certain limit, it may cause nuisance algal blooms, subsequently depletion of dissolved oxygen (DO), and fish kill, etc. (Edwards et al. 1990). Hence, proper design of an aquaculture farm is required for efficient use of the resources available in the wastewater.

In line with this, the current paper adopts a case study approach to demonstrate how a successful wastewater reuse scheme has been adopted by the Mudialy Fisherman's Cooperative Society in Kolkata, West Bengal, that has contributed to sustainable development. Using urban refuse and the polluted waters of the city, this society has developed a completely indigenous bioengineering system that can perform three important functions:

1. Improve industrial wastewater quality before releasing it in the River Ganges;
2. Use the wastewater as a habitat to grow fish; and
3. Develop an ecologically balanced system to accommodate a number of animals and plant species, thus increasing the biodiversity of the area.

The primary objectives of this study were to ascertain that conditions really are favorable for healthy aquaculture activities throughout the year and to bring the efforts of the MFCS into the limelight. This paper is an effort to encourage the local fishermen so that they get reinforcement and further boosts from the community and governmental and non-governmental organizations.

To achieve this target, a thorough survey of this wastewater aquaculture farm—as well as an assessment of the water quality in the farm by measuring different physicochemical parameters and the concentration of different heavy metals in the water tanks and ponds of the farm—was undertaken. It may be mentioned that physicochemical properties are considered a good monitor of water pollution since their concentrations are related to environmental pollution (Ashraf 2004). Moreover, as a shift in the community structure of plankton indicates water pollution, analysis of the plankton community at the MFCS ponds was also carried out.

MATERIALS AND METHODS

Description of the Study Area: Location and Climate

The Mudialy Fisherman's Cooperative Society (MFCS) is located in a densely populated area consuming 80 hectors about 10 km southwest from the heart of Kolkata, capital city of the state of West Bengal in Eastern India. It is based on about 80 hectares of wetland leased in from the Kolkata Port Trust (KPT). But through an injunction of the Honorable Kolkata High Court in response to an environment-loving NGO called Citizens' Forum, the *de facto* property rights to this resource have been vested with the poor fishermen of the MFCS since 1992. Moreover, this fishery is surrounded by a large number of industries. The MFCS converted the garbage dumping ground of Kolkata Port Trust into an aquaculture farm and have done a praiseworthy job of providing a source of income for poor farmers. The poor fishermen of MFCS have developed a unique system for recycling industrial and municipal waste (including solid waste and sewage), as well as developed a production center for horticulture, floriculture, animal husbandry, farm foresty, etc.

The climate in this wetland broadly resembles that of Kolkata, which, being located almost within a degree of the Tropic of Cancer, is at the limit of the torrid zone. With some variations, the temperature remains high throughout the year. The three major seasons are as follows:

1. Cold season: Sets in from the middle of November and lasts until the end of February. The mean temperature fluctuates between 15°C –20°C.
2. Hot season: Starts in March and lasts until mid-June, with mean temperatures between 30°C–40°C. Rainfall, which occurs more frequently toward the end of the season, is associated with thunder and lightning (Nor'westers).
3. Rainy season: Begins in mid-June and lasts until mid-September, stretching sometimes to October. The season is characterized by a high average temperature. Total rainfall during monsoons varies between 1,200 mm to 1,300 mm. Relative humidity varies between 80% and 85%.

The Technical System of MFCS

The average daily loading of sewage water is approximately 23 million liters, of which about 70% is from industries and the rest from domestic sewage (unpublished data collected from MFCS officials). This wastewater enters into the fishery where the technical system for treating the wastewater is located.

- The wastewater passes through a series of eight tanks known as anaerobic tanks, where the water is treated manually using liming or other biochemical purification techniques.
- Water hyacinths are usually kept near the anaerobic tanks to facilitate absorption of the oil, grease, and heavy metals in the effluent.
- Water flows from the first tank to the second through a narrow passage, which is a breeding ground for exotic fish that can survive the harsh conditions. These include omnivorous varieties such as *Tilapia nilotica* and air-breathing categories such as singi (*Heteropneustes fossilis*), magur (*Clarias batrachus*), and koi (*Anasbas testudineus*), which can endure even exotic stress.
- The water then flows into a third pond and so on through a saprophytic canal, with the water quality improving at each stage, therefore permitting better utilization of the recycled nutrients and mineral contents of the sewage for fish culture. Finally, the water is let into the Manikhal canal system, which eventually joins the Hooghly River.

Activities of the MFCS

Production and sale of fish is the core activity of MFCS. The society is also engaged in the sale of fishery inputs and consumer goods in the interests of its members and the community.

Species that are cultured include: Indian major carps (*Catla catla, labeo rohita, Cirrhinus mrigala,* and *Labeo bata*) and exotic carps (*Cyprinus carpio, Tilapia mossambica, Hypophthalmichthys molitrix, Ctenopharygodon idella,* and *Tilapia nilotica*).

As it is an environmental cooperative, a number of additional, peripheral, but ecologically and socially relevant environmental activities, such as plantations and construction of Nature Park, are also undertaken. According to a study conducted by the Zoological Survey of India in 1991, Nature Park contains as many as 120 varieties of bird fauna, of which 27 are migratory birds. The society is also successfully experimenting with promotion of spotted dear, ducks, and tortoise in the park. Member welfare activities, such as medical aid, education aid, old-age pension, etc., are also being pursued very vigorously by the society.

Sampling and Analysis of Water Samples

The wastewater is processed and treated as it passes through a number of ponds and tanks. Hence, the water quality analysis of the fishery was carried out by sampling water from different sampling sites, including:

1. All eight purification tanks as mentioned above; and
2. Five main inlet ponds for fish culture named 'Gorala,' 'Ghasbari,' 'Khudi,' 'Pond 9,' and 'Taltala.' Inlet ponds are meant for pisciculture (Figure 1).

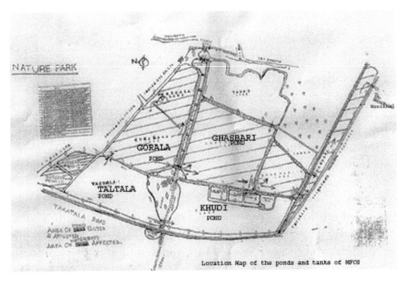

FIGURE 1 Location map for the ponds and tanks of MFSC (color figure available online).

Physicochemical properties analyzed included water pH, temperature, salinity, alkalinities, hardness, DO, BOD, COD, nitrate, phosphate, and sulfate. Apart from these parameters, concentrations of several heavy metals were also measured in both tanks and ponds of MFCS. The composition of the zooplankton community was analyzed only in the ponds as these are the actual sites for pisciculture. These analyses were carried out in different seasons, viz. winter (December–February), summer (March–May), and in the rainy season (July–August), to observe any seasonal variation in the parameters. All analyses were carried out following the standard methods of the American Public Health Association (APHA 1992).

RESULTS & DISCUSSION

The physicochemical characteristics of water in the eight tanks and five ponds analyzed during different seasons are shown in Tables 1 and 2, respectively. The concentrations of different heavy metals (measured by atomic absorption spectroscopy) in the tanks and ponds are given in Tables 4 and 5. Zooplankton compositions in the ponds of MFCS are shown in Table 5. It is evident from the results that the water-quality parameters in the ponds showed improvements after being successively purified in the tanks. Alkalinity and dissolved oxygen content are higher in ponds when compared to that of tanks, and this reflects a suitable environment for fish culture (Wurts & Durborow 1992; Boyd & Tucker 1998; Satyanarayana 2001). Salinity is an ecological factor of considerable importance influencing the types of organisms that live in a water body. The salinity of the ponds in MFCS remained well within the range for freshwater aquatic life (Brian 2003). Hardness, BOD, COD, sulphate, phosphate, and nitrate in ponds of MFCS

TABLE 1 The Physicochemical Characteristics of Water in the Eight Tanks

		Range of values in the eight tanks		
Sl. No.	Parameters	Dec. '08–Feb. '09	Mar.–May'09	June–Aug.'09
1.	Temperature	20.6°C–22.5 °C	26°C–37.5°C	36°C–34°C
2.	pH	7.03–8.15	7.2–8.24	7.55–7.66
3.	DO	3.2–6.13 mg/l	3.5–6.5 mg/l	3.1–4.3 mg/l
4.	CO_2	23.3–36 mg/l	20.9–49.5 mg/l	24.8–27.9 mg/l
5.	Hardness	149.3–169 mg/l	148.5–169.5 mg/l	120–142 mg/l
6.	Alkalinity	42.1–48.5 mg/l	43.5–48.05 mg/l	41.8–47.9 mg/l
7.	Salinity	0.9–3.1 ppt	1–2.9 ppt	1.7–3.5 ppt
8.	BOD	2.3–3.53 mg/l	2.5–3.55 mg/l	3.2–3.8 mg/l
9.	COD	14.3–22.6 mg/l	11.5–25 mg/l	28.6–36 mg/l
10.	Nitrate	0.324–0.35 mg/l	0.325–0.357 mg/l	0.336–0.365 mg/l
11.	Sulphate	6.43–7.31 mg/l	6.4–7.3 mg/l	6.73–7.36 mg/l
12.	Phosphate	0.173–0.199 mg/l	0.181–0.196 mg/l	0.178–0.198 mg/l

TABLE 2 The Physicochemical Characteristics of Water in the Five Ponds

		Range of values in the five ponds		
Sl. No.	Parameters	Dec. '08–Feb. '09	Mar.–May '09	June–Aug. '09
1.	Temperature	20.2°C–22°C	25°C–36°C	37°C–34°C
1	pH	7.13–8.67	7.9–8.73	7.81–8.23
2.	DO	8.43–9.3 mg/l	8.35–9.6 mg/l	5.4–6.4 mg/l
3.	CO_2	10.6–12.2 mg/l	11.5–12.35 mg/l	11.7–14.8 mg/l
4.	Hardness	108.3–141 mg/l	110.5–142.5 mg/l	109–132 mg/l
5.	Alkalinity	31.6–36.8 mg/l	32–36 mg/l	31.5–37.3 mg/l
6.	Salinity	0.37–0.48 ppt	0.375–0.49 ppt	0.45–0.53 ppt
7.	BOD	1.83–2.06 mg/l	1.9–2.1 mg/l	3.6–4.3 mg/l
8.	COD	53.6–64.6 mg/l	59.5–63.5 mg/l	39.3–44.6 mg/l
9.	Nitrate	0.305–0.32 mg/l	0.305–0.335 mg/l	0.35–0.41 mg/l
10.	Sulphate	4.2–5.7 mg/l	4.3–5.6 mg/l	4.3–5.6 mg/l
11.	Phosphate	0.038–0.074 mg/l	0.037–0.071 mg/l	0.037–0.073 mg/l

TABLE 3 Concentrations of Different Heavy Metals in the Tanks

		Range of values in the eight tanks		
Sl. No.	Heavy Metals	Dec. '08–Feb. '09	Mar.–May '09	June–Aug. '09;
1	Fe	0.32–0.41 mg/l	0.34–0.46 mg/l	0.36–0.45 mg/l
2.	Cu	0.048–0.062 mg/l	0.048–0.065 mg/l	0.0046–0.02 mg/l
3.	Pb	0.036–0.056 mg/l	0.018–0.058 mg/l	0.025–0.073 mg/l
4.	Cd	0.038–0.051 mg/	0.040–0.05 mg/l	0.043–0.056 mg/l
5.	Zn	0.228–0.307 mg/l	0.177–0.328 mg/l	0.277–0.428 mg/l

TABLE 4 Concentrations of Different Heavy Metals in the Ponds

		Range of values in the five ponds		
Sl. No.	Heavy Metals	Dec. '08–Feb. '09	Mar.–May '09	June–Aug. '09
1	Fe	0.29–0.35 mg/l	0.22–0.28 mg/l	0.27–0.32 mg/l
2.	Cu	0.024–0.057 mg/l	0.029–0.06 mg/l	0.0014–0.0064 mg/l
3.	Pb	0.042–0.054 mg/l	0.016–0.044 mg/l	0.035–0.061 mg/l
4.	Cd	0.023–0.034 mg/l	0.031–0.04 mg/l	0.031–0.042 mg/l
5.	Zn	0.107–0.187 mg/l	0.121–0.206 mg/l	0.212–0.308 mg/l

remained within the permissible range for pisciculture during the whole study period.

Moreover, the ponds of the MFCS exhibited zooplankton in great abundance (in number and diversity) throughout the water column, and this may be one of the reasons that MFCS is running their fishery proficiently to date. This plankton diversity can be correlated to higher alkalinity, which is thought to be conducive for higher growth of plankton. This, in turn, increases the oxygen concentration and nutrient turnover. It is known that plankton abundance and distribution are strongly dependent on factors such

TABLE 5 Zooplankton Composition

Sl. No.	Zooplanktons	Approximate plankton numbers/50c.c. in the five ponds		
		Dec. '08–Feb. '09	Mar.–May '09	June–Aug. '09
1.	*Moina*	22	20	27
2.	*Mesocyclops hyalinus*	61	49	74
3.	*Daphnia*	9	9	8
4.	*Ceriodaphnia*	14	11	10
5.	*Allodiaptomus*	22	22	14
6.	*Filinia*	8	3	4
7.	*Asplanchna*	3	1	5
8.	*Mesocyclops leukarti*	58	52	63
9.	*Diaphnosoma*	5	5	7
10.	*Heliodiaptomus*	4	3	4

as ambient nutrient concentration, the physical state of water column, and the abundance of other plankton (Boyd & Ahmad 1987; Satyanarayana 2001; Jensen & Blankston 1988). No seasonal variation, i.e., winter and early summer, regarding abundance of plankton was observed from this study. Concentrations of heavy metals (Fe, Cu, Pb, Zn, Cd) in the present investigation were reasonably lower than the objectionable level from the point of view of pisciculture (Stone & Thomforde 1977).

From the ongoing discussions, it is very evident that all the ponds and tanks of MFCS maintain an environment quite favorable for aquaculture practices. Even the rainy season (June–August), when water bodies tend to receive large amounts of industrial and other terrestrial runoffs, did not influence the physicochemical properties of water in the ponds and tanks, probably owing to the efficient purification system employed in the MFCS. Thus, MFCS is successfully thriving on the poison constantly being released in the form of dangerous industrial effluents from the sewage pipes of a large number of industries located in this area.

CONCLUSION

As described, it is quite clear that MFCS has succeeded in maintaining water quality parameters within the acceptable range for healthy aquaculture. Mudialy Fishery has done a commendable job of converting the domestic waste and industrial effluent into a highly structured, proficient, commercial production center for vegetable and fish. In the process the poor fishermen of MFCS have used the wetlands for wastewater fisheries by cleaning the water and converting the area into a 'smiling' Nature Park, as it is known today. This demonstrates the versatility and enormous productivity of wetlands in the Indo-Gangetic plains, which, when managed by the most intimate stakeholders—the fishermen with sustainable technology and a suitable organizational format—are capable of producing better results for society.

However, it is important to point out that, in spite of such an incredible achievement, the MFCS is not without problems. This fishery has faced sporadic incidences of fish kills in the past, which has been a major setback for the poor fishermen. Another major concern for MFCS is that it has never been able to attain good grade in audit classification—a fact that undermines its achievements to some extent. It should be in the immediate interests of MFCS to hire good scientific persons as heir advisors and efficient auditors who can train the fishermen to keep the accounts in a better manner.

Unfortunately, most states, academic bodies, banks, and NGOs are not even aware of MFCS, nor do the members or officials associated with the Mudialy experiment have an opportunity to take it further through learning from others. This article is an effort to acknowledge the poor fishermen of MFCS and to throw some light on the incredible achievements of MFCS so that such a profitable 'environmental cooperative society' receives both public and governmental support.

The MFCS has created a very important niche for itself by providing economic relief to the poor fishermen as well as the conservation of biodiversity. Thus, it is in the interest of not only the local fisherman but also general citizens that the MFCS continues to functions properly.

REFERENCES

American Public Health Association (APHA). 1992. *Standard methods for the examination of water and wastewater*, 18th ed. Washington, DC: American Public Health Association.

Ashraf, W. 2004. Levels of selected heavy metals in tuna fish. *The Arabian Journal for Science and Engineering* 31 (IA): 89–92.

Boyd, C.E, and T. Ahmad. 1987. *Evaluation of aerators for channel catfish farming.* Bulletin no. 584. Auburn, AL: Alabama Agricultural Experiment Station, Auburn University.

Boyd, C.E., and C. S. Tucker. 1998. *Pond aquaculture water quality management.* Norwell, MA: Kluwer Academic Publisher. ISBN-10:0412071819.

Brian, D. R. M. 2003. Ecologically sustainable water management, managing river flows for ecological integrity. *Ecological Application* 13(1): 206–224.

Central Institute of Freshwater Aquaculture (CIFA). 1998. *Sewage treatment through aquaculture.* Bhubaneswar, India: CIFA.

Edwards, P., and R. S. V. Pullin. 1990. Wastewater-fed aquaculture. Proceedings of the International seminar on wastewater reclamation and reuse for aquaculture, Calcutta, India, 6–9 December 1988, Environmental Sanitation Information Center, Asian Institute of Technology, Bangkok.

Jensen, G.I, and J. D. Blankston. 1988. *Guide to oxygen management and aeration in commercial ponds.* Baton Rouge, LA: Louisiana Agricultural Experiment Station. Louisiana Cooperative Extension Service.

Satyanarayan, K. 2001. *Element analysis: A vital tool for water quality assessment, bio-monitoring and zooplankton diversity*. Edited by B. K. Sharma. New Delhi, India: Ministry of Environment and Forest, Govt. of India.

Stone, M.N., and K. H. Thomforde. 1977. *Understanding your fish pond water analysis report*. Pine Bluff, AR: Cooperative Extension Program, University of Arkansas at Pine Bluff.

Wurts, W. A., and R. M. Durborow. 1992. *Interactions of pH, carbon dioxide, alkalinity and hardness in fish ponds. Fact sheet no. 464.* Stoneville, MS: Southern Regional Aquaculture Center. http://www.msstate.edu/dept/srac/fslist.htm

The Effect of Partial Replacement of Dietary Fishmeal with Fermented Prawn Waste Liquor on Juvenile Sea Bass Growth

N. M. NOR[1], Z. ZAKARIA[1], M. S. A. MANAF[2], and M. M. SALLEH[3]

[1]*Department of Chemistry, Faculty of Science, Universiti Teknologi Malaysia, Johor, Malaysia*
[2]*Brackishwater Aquaculture Research Centre, Gelang Patah, Johor, Malaysia*
[3]*Faculty of Bioscience and Bioengineering, Universiti Teknologi Malaysia, Johor, Malaysia*

A feeding trial was conducted for 49 days to evaluate the effect of partially substituting fishmeal with fermented prawn waste liquor (FPWL) in juvenile sea bass diets at 10%, 20%, and 30% of the total diet. Growth performance of sea bass from 16 g up to 40 g fed with FPWL-supplemented diet was not significantly different from the all fishmeal control diet. The most cost-effective diet included FPWL at 30%, with weight gain, feed conversion ratio, and protein efficiency ratio of 180%, 1.78%, and 1.2% respectively.

INTRODUCTION

Agriculture and fisheries by-products and non-food grade feed materials have been identified as alternatives to reduce dependency on expensive fishmeal in aquaculture diets (Food and Agriculture Organization 2000). One of the potential protein sources abundantly available in Malaysia is prawn processing waste. Prawn waste contains 35–55% calcium, 15–40% protein, and 14–30% chitin (Ornum 1992; Legaretta, Zakaria, & Hall 1996; Zakaria, Sharma, & Hall 1998). Conventionally, chitin has been recovered from prawn waste by chemical methods, the protein fraction being discarded in the process. In contrast, microbial lactic acid fermentation improves protein recovery and reduces the amount of chemicals used (Zakaria, Sharma, & Hall 1998). The fermented prawn silage thus produced is rich in protein and can be used in animal feeds.

The use of fermented silage in livestock feeds and aquafeeds has been widely studied. Fagbenro, Jauncey, and Haylor (1994) reported that fermented fish silage and soybean meal could replace up to 75% of fishmeal in diets fed to juvenile *Oreochromis niloticus* and *Clarias gariepinus*. Espe, Haaland, and Njaa (1992) and Heras, Mcleod, and Ackman (1994) found that fish silage in salmon (*Salmo salar*) diets had no affect on growth, but significantly reduces costs. Nwanna (2003) reported that prawn waste silage was a good alternative protein source for *C. gariepinus*. In this study, prawn waste silage was incorporated in sea bass diets to partially replace fishmeal, and the effect on growth performance was determined in a 49-day feeding trial.

MATERIALS AND METHODS

Frozen tiger prawn (*Penaeus monodon*) wastes (heads, exoskeletons, and tails) were obtained from a local seafood processing factory, minced, and kept frozen until used. MRS (De Man, Rogosa and Sharpe) broth (Merck, Germany) was used to culture the bacterium, *Lactobacillus sp.* in the laboratory overnight. To thawed minced prawn waste was added 10% (by wet weight) glucose (MERCK, Germany) and 10% (by volume) bacteria culture and incubated at 37°C. The container was covered to provide anaerobic conditions and stirring was done occasionally, especially during the first 24 hours, to make the fermentation mixture homogeneous and prevent spoilage at the top of the mixture. Fermentation went on for three days, after which the fermented slurry was filtered to separate the fibrous solid (chitin) from the proteinaceous fermented prawn waste liquor (FPWL). The FPWL was then mixed with other dietary ingredient (Table 1) in various percentages to produce FPWL-based fish diets (Table 2). Diet 1 was the control diet, which

TABLE 1 Proximate Analysis of Ingredients (%, w/w) used in the Fish Diet Formulations

	Protein	Fat	Ash	Chitin	Fiber	NFE[1]
Fishmeal	68.0	10.8	13.7	n.d	0.5	7.0
Soya bean meal	47.6	0.3	8.6	n.d	7.1	36.4
Corn flour	0.3	0.2	0.9	n.d	1.7	96.9
Fish oil	0	100	0	n.d	0	0
Fermented prawn waste liquor	47.3	6.1	16.1	6.5	0.8	29.7

[1]Nitrogen-free extracts [100 - (protein + fat + ash + fiber)].
[2]n.d = not determined.

TABLE 2 Percentage Inclusions of Ingredients in the Fish Diet Formulations

	Diets			
Ingredients	1	2	3	4
Fishmeal	39.7	32.8	25.8	18.8
Soy bean meal	37.8	37.8	37.8	37.8
Fermented prawn waste liquor	0	10.0	20.0	30.0
Corn flour	14.2	11.0	7.8	4.7
Fish oil	5.6	5.7	5.9	6.0
Others[1]	2.7	2.7	2.7	2.7

[1] Others = 30 mg kg^{-1} diet of coated vitamin C; 50 mg kg^{-1} diet of L-lysine; 50 mg kg^{-1} diet of L-methionine; 1.0% choline chloride; 0.5% titanium dioxide; 1.0% dicalcium phosphate; and 0.2% of premix.

TABLE 3 Proximate Analyses of Formulated Fish Diets

Fish Diet	Percentage, % (w/w)					ME^2(kJ/g)
	Protein	Fat	Ash	Fiber	NFE^1	
1	46.1 ± 0.1	9.5 ± 0.1	9.5 ± 0.02	3.59 ± 0.2	31.4	16.51
2	46.2 ± 2.4	11.2 ± 1.8	9.9 ± 0.00	3.40 ± 0.1	34.0	16.83
3	46.0 ± 5.6	11.1 ± 2.6	10.2 ± 0.08	2.69 ± 0.1	30.0	16.88
4	48.4 ± 0.3	13.1 ± 0.4	10.8 ± 0.00	1.89 ± 0.5	26.0	17.33

[1] Nitrogen-free extract [100 - (protein + fat + ash + fiber)].
[2] Metabolized energy. Fish diet 1, 2, 3, and 4 contain 0 %, 10%, 20%, and 30% FPWL respectively as in Table 2.

contained no FPWL. Diet 2 contained 10% FPWL, Diet 3, 20% FPWL, and Diet 4, 30% FPWL. The test ingredients and formulated diet were analyzed for proximate components using AOAC (2002) methods (Table 3).

Sea bass juveniles (initial length 7 cm) were purchased locally and stocked into16, 150-L fibreglass tanks at the Brackishwater Aquaculture Research Centre in Southern Malaysia. Each tank was supplied with a continuous flow of filtered aerated brackish water. Sea bass juveniles were acclimatized to experimental conditions for two weeks, after which time fish were starved for 48 hours and individually weighed. Sea bass juveniles with an initial weight averaged 16.2 ± 0.4 g were chosen and randomly redistributed into the 16 tanks at 10 fishes/tank (four replicates/treatment). The fish were hand-fed thrice daily for 49 days. All fish were weighed once weekly, and mean weight data were used to assess the growth performance. Fish stress during sampling was minimized by anaesthetizing the fish with 200 ppm of 2-phenoxyethanol (Sigma, United States) (Mohammed-Suhaimee, Zakaria, & Ng 2006). At the termination of the experiment, the fish were starved for 24 hours before harvest, killed with an overdose of 2-phenoxyethanol, and weighed for growth performance analysis. Feed intake per replicate during the period was also recorded.

Data were analyzed with one-way analysis of variance (ANOVA) using SPSS 9. Duncan's multiple range tests was then used to compare differences among treatment means. Differences were considered significant at P < 0.05. Simple cost analysis was carried out to make a comparison between the control diet and the prawn waste diet.

RESULTS AND DISCUSSION

Commercial fishmeal contains 68% protein and 0.5% fiber, while FPWL contains 47.3% protein and 0.8% fiber. Nwanna (2003) reported that fermented shrimp head meal contained 58.96% crude protein. This difference is probably due to the fact that Nwanna (2003) used only prawn heads, while we used total prawn wastes including the head, exoskeleton, and tail. In any case, all of the diets met the nutrient requirements of sea bass: 40–50% protein, 10% crude lipid, and metabolized energy of about 17 kJ/g (Cuzon, Chou, & Fuchs 1989; Sakaras et al. 1988, 1989; Catacutan & Coloso, 1995; Williams & Barlow, 1999, 2003). Crude fiber content of each diet (2–4%) is lower than those used by Nwanna (2003), which was 4–5%, and Fanimo et al. (2004), which was 8–9%.

Throughout the trial, water quality in the tanks was within acceptable ranges for sea bass culture (Rosly Hassan 1990): dissolved oxygen of 5.01–6.4 mg/L, pH of 7.4–7.8, temperature of 27°–29°C, and salinity of 22–30 ppt.

Fish fed with FPWL-based diets showed good growth performance (Figure 1, Table 4). Fish in the control diet treatment performed the best, but the difference was not significant (P < 0.05). Among the three FPWL-based diets, 30% inclusion of FPWL was closest to the control diet. SGR

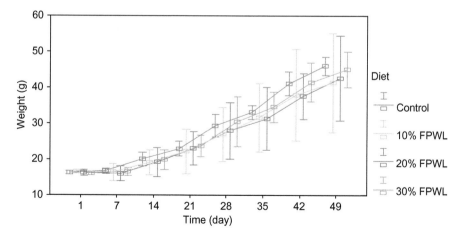

FIGURE 1 Weight gain of sea bass fed with diets containing various levels of fermented prawn waste liquor (FPWL) for 49 days.

TABLE 4 Growth, Specific Growth Rate (SGR), Feed Conversion Ratio (FCR), and Protein Efficiency Ratio (PER) of Sea Bass Fed with Various Inclusion Rates of Fermented Prawn Waste Liquor (FPWL)

Formulated Diet	Weight gain (%)	SGR (%/day)	FCR	PER
1 (0% FPWL)	181.92 ± 1.15[a]	2.11 ± 0.02[a]	1.85 ± 0.54[a]	1.18 ± 0.08[a]
2 (10% FPWL)	148.22 ± 7.13[a]	1.82 ± 0.15[a]	2.90 ± 0.22[a]	0.93 ± 0.29[a]
3 (20% FPWL)	162.88 ± 6.65[a]	1.95 ± 0.14[a]	2.84 ± 0.17[a]	0.80 ± 0.23[a]
4 (30% FPWL)	179.82 ± 2.76[a]	2.10 ± 0.15[a]	1.77 ± 0.40[a]	1.18 ± 0.12[a]

[a] Means with the same letter were not significantly different.

values of 1.8 to 2.1 were consistent with the results of Cuzon, Chou, and Fuchs (1989), and Mohammed-Suhaimee, Zakaria, and Ng (2006), in which sea bass fed with diets containing 40–45% protein gave SGR values of 1.5–2.1. PER values in this study are comparable with a PER of 1.6 obtained by Mohammed-Suhaimee, Zakaria, and Ng (2006). The feed conversion ratio (FCR) values recorded in this study were 1.8–2.9. These values were different from the 1.0–1.5 reported by Catacutan and Coloso (1995), Williams and Barlow (1999, 2003), and Bautista (1994).

The two main ingredients involved in the fermentation of prawn waste were glucose and MRS broth used for growth of bacteria. The total cost to produce 1 kg of fermented prawn waste liquor (FPWL) was US$0.65/kg, 44% cheaper than fishmeal (US$1.17/kg). The cost of all ingredients used in the diet formulation is listed in Table 5, and the costs of the diets were calculated based on those numbers (Table 6). The control diet was the most expensive, and the 30% FPWL diet was the least expensive (7.7% cheaper than the control diet).

TABLE 5 Cost for Each Ingredient used in the Experimental Diet Formulation in 2009

Ingredients	Cost/ kg (US$)*
Fishmeal	1.17
Soybean meal	0.41
Fermented prawn waste liquor	0.65
Corn flour	0.21
Fish oil	0.94
Vitamin mix	24.29
Vitamin C	114
Binder	1.86
Choline	1.43
Mineral mix	24.30

TABLE 6 The Cost of Various Formulated Diets

Item	Diet 1		Diet 2		Diet 3		Diet 4	
	kg	Cost (US$)	kg	Cost (US$)	kg	Cost (US$)	kg	Cost (US$)
FM	3.97	4.64	3.30	3.86	2.60	3.04	1.88	2.20
SBM	3.78	1.55	3.80	1.55	3.80	1.55	3.78	1.55
FPWL	0.00	0	1.00	0.65	2.00	1.30	3.0	1.95
Corn flour	1.42	0.30	1.10	0.23	0.78	0.16	0.47	0.10
Fish oil	0.56	0.53	0.56	0.53	0.56	0.53	0.56	0.53
Vitamin mix	0.02	0.49	0.02	0.49	0.02	0.49	0.02	0.49
Vitamin C	0.0003	0.03	0.0003	0.03	0.0003	0.03	0.0003	0.03
Masterqube	0.10	0.19	0.10	0.19	0.10	0.19	0.10	0.19
Choline	0.10	0.14	0.10	0.14	0.10	0.14	0.10	0.14
Mineral mix	0.0497	1.27	0.0497	1.21	0.0497	1.21	0.0497	1.21

FM = fishmeal; SBM = soybean meal; and FPWL = fermented prawn waste liquor.

REFERENCES

Association of Official Analytical Chemists (AOAC). 2002. *Official methods of analysis*. 13th ed. Edited by W. Horwitz. Washington, DC: AOAC.

Bautista, M.N. 1994. *Feeds and feeding of milkfish, Nile tilapia, Asian sea bass, and tiger shrimp*. Iloilo, Philippines: SEAFDEC.

Catacutan, M.R., and R.M. Coloso. 1995. Effect of dietary protein to energy ratios on growth, survival, and body composition of juvenile Asian sea bass, *Lates calcarifer*. *Aquacult*. 131:125–133.

Cuzon, G., R. Chou, and J. Fuchs. 1989. Nutrition of the sea bass *Lates calcarifer*. In *Advances in tropical aquaculture, Workshop in Tahiti, French Polynesia, February 20-March 4, 1989*, Eds. J. Barret, J. Calvas, G. Cuzon, J. Fuchs, and M. Weppe, 757–763.

Espe, M., H. Haaland, and L.R. Njaa. 1992. Autolyzed fish silage a feed ingredient for Atlantic Salmon (*Salmo salar*). *Comp. Biochem. Physiol.* 103:369–372.

Fagbenro, O., K. Jauncey, and G. Haylor. 1994. Nutritive value of diet containing dried lactic acid fermented fish silage and soybean meal for juvenile *Oreochromis niloticus* and *Clanas gariepinus*. *Aquat. Living Resour.* 7:79–85.

Fanimo A.O., B.O. Odugawa, O.O. Odugawa, O.Y. Ajasa, and O. Jegede. 2004. Feeding value of shrimp meal for growing pigs. *Arch. Zootec.* 53:77–85.

Food and Agriculture Organization (FAO). 2000. The Bangkok declaration and strategy conference on aquaculture in third millennium. Bangkok: FAO.

Heras, H., C.A. Mcleod, and R.G. Ackman. 1994. Atlantic dogfish silage vs. herring silage in diets for Atlantic salmon (*Salmo salar*): Growth and sensory evaluation of fillets. *Aquacult.* 125:93–103.

Legaretta, G.I., Z. Zakaria, and G.M. Hall. 1996. Lactic acid fermentation of prawn waste: Comparison of commercial and isolated starter culture. In *Advances in chitin science*, Eds. A. Domard, C. Jeuniaux, R.A.A. Muzarelli, and G.A.F. Roberts, 396–406. Lyon, France: Jacques Andre.

Mohammed-Suhaimee, A.M., Z. Zakaria, and W.K. Ng. 2006. The Inclusion of urea-treated palm kernel meal as a feed ingredient in the diet of juvenile sea bass (*Lates Calcarifer*) affects growth performance. *J. Aqua Trop.* 21:59–76.

Nwanna, L.C. 2003. Nutritional value and digestibility of fermented shrimp head waste meal by African catfish, *Clarias gariepinus*. *Pak. J. Nutr.* 2(6): 339–345.

Ornum, J.V. 1992. Shrimp waste—must it be wasted? *INFOFISH Int.* 6:48–52

Bin Rosly, H. 1990. Garispanduan Mutu Air Untuk Ternakan Ikan dan Udang Laut. Risalah Perikanan. Bil. 43. [Water quality standards for fish and prawn culture]. Trans. By Nor Hazurina Binti Haji Othman. Persekutuan, Malaysia: Department of Fisheries, Malaysia. Available at http://fri.gov.my/portal/portalimages/file_282.PDF.

Sakaras, W., M. Boonyaratpalin, N. Unpraser, and P. Kumpang. 1988. *Optimum dietary protein energy ratio in sea bass feed I. Technical Paper. No.* 7. Thailand: Rayong Brackishwater Fisheries Station.

Sakaras, W., M. Boonyaratpalin, N. Unpraser, and P. Kumpang. 1989. *Optimum dietary protein energy ratio in sea bass feed II. Technical Paper. No.* 8. Thailand: Rayong Brackishwater Fisheries Station.

Williams, K.C., and C.G. Barlow. 1999. *Dietary requirement and optimal feeding practices for barramundi (Lates calcarifer)*. Project 92/63, Final report to Fisheries R&D Corporation, Canberra, Australia.

Williams, K.C., C.G. Barlow, L. Rodgers, I. Hockings, C. Agcopra, and I. Ruscoe. 2003. Asian sea bass (*Lates calcarifer*) perform well when fed pellet diets high in protein and lipid. *Aquacult.* 225:191–206.

Zakaria. Z., G. Sharma, and G.M. Hall. 1998. Lactic acid fermentation of scampi waste in a rotating bioreactor for chitin recovery. *Process Biochem.* 33:1–6.

Growth Performance and Resistance to *Streptococcus iniae* of Juvenile Nile Tilapia (*Oreochromis niloticus*) Fed Diets Supplemented with GroBiotic-A and Brewtech Dried Brewers Yeast

KUNTHIKA VECHKLANG[1], CHHORN LIM[2], SURINTORN BOONANUNTANASARN[3], THOMAS WELKER[2], SAMORN PONCHUNCHUWONG[3], PHILLIP H. KLESIUS[2], and CHOKCHAI WANAPU[1]

[1] *School of Biotechnology, Institute of Agricultural Technology, Suranaree University of Technology, Nakhon Ratchasima, Thailand*
[2] *Aquatic Animal Health Research Laboratory, USDA-ARS, MSA, Auburn, Alabama, United States*
[3] *School of Animal Production Technology, Institute of Agricultural Technology, Suranaree University of Technology, Nakhon Ratchasima, Thailand*

This study was conducted to evaluate the effect of dietary levels of Brewtech dried brewers yeast (BY) and GroBiotic-A (GB) on growth performance, proximate body composition, immune response, and resistance of juvenile Nile tilapia to Streptococcus iniae *challenge. A practical basal (control) diet formulated to contain approximately 32% crude protein and 6% lipid was supplemented with 1% and 2% of BY or GB. Each diet was fed to Nile tilapia in quadruplicate aquaria for 12 weeks. Weight gain, feed intake, survival, and whole body proximate composition of*

The authors would like to thank Suranaree University of Technology, Rajamangala University of Technology Isan, and the Aquatic Animal Health Research Unit, USDA-ARS, MSA, for funding support of this project; International Ingredient Corporation (St. Louis, MO, USA) for donation of GroBiotic-A and Brewtech dried brewers yeast; and Ms. Rashida Eljack, Aquatic Animal Health Research Unit, USDA ARS, Auburn, Alabama, for technical assistance.

fish were not significantly affected by dietary treatments. Serum total protein, total immunoglobulin, lysozyme, and agglutinating antibody titer to S. iniae *were not significantly affected by dietary supplementation of BY or GB. However, serum haemolytic complement activity (SH50) of fish fed 1% BY was significantly higher than those of fish fed the control diet and diets supplemented with 2% BY or GB. Cumulative mortality of fish 20 days post-challenge with* S. iniae *was unaffected by dietary treatments. However, fish fed diets supplemented with 1% BY and 2% BY or GB had substantially reduced and earlier cessation of mortality.*

INTRODUCTION

Tilapia—because of their fast growth, enormous adaptability to a wide range of physical and environmental conditions, ability to reproduce in captivity, resistance to handling and disease, good flesh quality, feed on a low trophic level, and excellent growth rate on a wide variety of natural and artificial diets—are the most successfully cultured species worldwide (Lim & Webster 2006). They are presently cultured in virtually all types of production systems, in both fresh and salt water, and in tropical, subtropical, and temperate climates. They are increasingly recognized as the species of choice for intensive aquaculture and are likely to become the most important cultured fish in the world (Fitzsimmons 2006). According to the American Tilapia Association, global farm-raised tilapia production is expected to reach 3 million metric tons by 2010, compared to 2.6 million metric tons in 2007.

However, one major problem associated with intensive fish culture is the increased susceptibility of fish to infectious diseases, including streptococcal disease in tilapia caused by *Streptococcus iniae*. The problem of streptococcal disease is worldwide (Muzsquiz et al. 1999), and the annual loss to the aquaculture industry is estimated to be over $100 million (Shoemaker et al. 2001). Commercially, antibiotics have been supplemented in aqua feeds for treatment and prevention of bacterial disease of aquatic animals (Li & Gatlin 2005). The use of antibiotics can lead to the emergence of antibiotic-resistant bacteria, and contamination in food products and the environment (Food and Agriculture Organization of the Untied Nations [FAO] 2002). The use of antibiotics in animal production has been banned in EU countries and is increasingly under public scrutiny and criticism in most other countries. Consequently, a wide variety of products ranging from polysaccharides, plant extracts, and some nutrients have been added in fish diets as immunostimulants to stimulate immune system function and/or their resistance to infectious diseases, or serve as adjuvant to improve vaccine efficacy (Sakai 1999; Gannam & Schrock 2001).

Brewtech dried brewers yeast (BY), *Saccharomyces cerevisiae*, is a natural product of the brewing industry containing various immunostimulating compounds such as β-glucans, nucleic acids, and mannans. It has been shown to positively influence non-specific immune responses as well as growth of various fish species and thus may serve as an excellent health promoter for fish culture (Li & Gatlin 2003, 2004, 2005; Waszkiewicz-Robak & Karwowska 2004). GroBiotic-A (GB), a commercial prebiotic consisting of a mixture of partially autolyzed brewers yeast, dairy ingredient components, and dried fermentation products, has been shown to enhance the growth performance, feed efficiency, and survival of hybrid striped bass to *S. iniae* and *Mycobacterium marinum* (Li & Gatlin 2004, 2005).

Thus, the present study was conducted to evaluate the effects of feeding diets supplemented with BY and prebiotic GB on growth performance, proximate body composition, immune response, and resistance of juvenile Nile tilapia (*Oreochromis niloticus*) to *S. iniae* challenge.

MATERIALS AND METHODS

Experimental Fish and Husbandry

Juvenile Nile tilapia spawned and reared at our laboratory on commercial fry and fingerling diets were acclimated to laboratory conditions and fed the basal experimental diet without BY or GB supplementation for two weeks. At the end of the acclimation period, fish with an average weight of 13.35 ± 0.11 g (mean $\pm$ SEM) were randomly selected and stocked in 20, 57-L glass aquaria at a density of 30 fish/aquarium. The aquaria were supplied with flow-through dechlorinated city water at an initial rate of about 0.6 L/min and increased gradually to about 1.0 L/min by the sixth week of the trial. Water flow rates were checked and adjusted twice daily to ensure proper water exchange. Water temperature was maintained constant ($26 \pm 1°C$) by a centralized water heater. The water was continuously aerated with air stones, and the photoperiod was maintained on a 12:12-h light: dark schedule. Dissolved oxygen and temperature in three randomly chosen aquaria were measured once every other day using an YSI model 58 Oxygen Meter (Yellow Spring Instrument, Yellow Spring, OH, USA). During the trial, water temperature and dissolved oxygen averaged $25.27 \pm 0.10°C$ and 5.34 ± 0.06 mg/L, respectively.

Experimental Diets and Feeding

A practical basal diet was formulated to contain approximately 32% crude protein, 6.0% crude lipid, and 2,900 kcal of digestible energy (DE)/kg based on feedstuff values reported in NRC (1993). Brewtech dried brewers yeast (BY) and GroBiotic-A (GB) provided by International Ingredient Corporation, St. Louis, MO, was added to the basal diets at levels of 0%, 1%, and 2% (Table 1). The levels of soybean meal, corn oil, and celufil (non-nutritive filler) were adjusted to maintain equal levels of dietary protein and lipid. All diets were supplemented with vitamins and minerals in amounts to meet the known requirements of tilapia (Lim & Webster 2006). Dry ingredients were thoroughly mixed for 10 min in a Hobart mixer (Hobart Corporation, Troy, OH, USA) before the oil was added. After the oil was dispersed, approximately 280 mL of deionized water/kg of diet was added. The moist mixture was extruded through a 3-mm diameter die in a Hobart meat grinder. The resulting moist pellets were air-dried at room

TABLE 1 Ingredients and Proximate Composition (%) of Five Experimental Diets

Ingredient	Percent (%) in diets				
	1	2	3	4	5
Menhaden fish meal	8.00	8.00	8.00	8.00	8.00
Soybean meal	45.00	44.10	43.30	44.30	43.60
Corn meal	23.50	23.50	23.50	23.50	23.50
Wheat middling	13.60	13.60	13.60	13.60	13.60
Corn Oil	3.50	3.45	3.40	3.45	3.40
Carboxymethyl cellulose	3.00	3.00	3.00	3.00	3.00
Dicalcium phosphate	1.00	1.00	1.00	1.00	1.00
Vitamin premix[1]	0.50	0.50	0.50	0.50	0.50
Mineral premix[2]	0.50	0.50	0.50	0.50	0.50
Brewers yeast	0.00	1.00	2.00	0.00	0.00
GroBiotic-A	0.00	0.00	0.00	1.00	2.00
Celufil	1.40	1.35	1.20	1.15	0.90
Total	100.00	100.00	100.00	100.00	100.00
Ethoxyquin	(0.02% or 200 mg/kg diet)				
Determined nutrition content (% as is)					
Dry matter	90.51	90.89	90.70	89.37	89.13
Protein	32.32	332.5	332.23	32.42	33.73
Fat	6.04	6.17	6.09	6.36	6.31
Ash	6.95	6.76	6.80	6.66	6.75
DE (kcal/kg diet)[3]	2,900	2,900	2,900	2,900	2,900

[1]Vitamin premix, diluted in cellulose, provided by the following vitamins (mg/kg diet): vitamin A (retinyl acetate), 4,000 IU; vitamin D_3 (cholecalciferol), 2,000 IU; vitamin K (menadione sodium bisulfide), 10; vitamin E (α-tocopheryl acetate), 50; thiamin hydrochloride, 10; riboflavin, 12; pyridoxine hydrochloride, 10; D-calcium pantothenate, 32; nicotinic acid, 80; folic acid, 2; vitamin B_{12}, 0.01; biotin, 0.2; choline chloride, 400; and vitamin C (as L-ascorbyl-2-polyphosphate, 45% vitamin C activity), 60.

[2]Trace mineral premix provided by the following minerals (mg/kg diet): zinc (as $ZnSO_4.7H_2O$), 150; iron (as $FeSO_4.7H_2O$), 40; manganese (as $MnSO_4.7H_2O$), 25; copper (as $CuCl_2$), 3; iodine (as KI), 5; cobalt (as $CoCl_2.6H_2O$), 0.05; and selenium (as Na_2SeO_3), 0.09.

[3]DE (digestible energy) was calculated based on feedstuff values reported in NRC (1993).

temperature (24°C) to a moisture content of about 10%. Pellets were ground into small pieces, sieved to obtain appropriate sizes, and stored frozen in plastic bags at −20°C until fed (Peres et al. 2003). Proximate composition of the experimental diets is given in Table 1.

Fish in four randomly assigned aquaria were fed one of the five experimental diets twice daily (between 07.30–08.30 and 15.00–16.00 h) to apparent satiation for 12 weeks. The amount of diet consumed was recorded daily by calculating the differences in weight of diets prior to the first and after the last feeding. Once a week, aquaria were scrubbed and accumulated waste was siphoned. On cleaning days, fish were fed only in the afternoon. Feed was not offered on sampling days.

Growth Measurements

Fish in each aquarium were removed, anesthetized with 150 mg/L tricaine methanesulfonate (MS-222; Argent Chemical Laboratories, Redmond, WA, USA), counted, and group weighed every three weeks, following 16 h of feed deprivation. When fish were removed for weighing, aquaria were cleaned thoroughly, and three-fourths of the water drained. Weight measurements and fish counts were used for estimation of weight gain, feed efficiency ratio (FER; wet weight gain/dry feed intake), and survival.

Proximate Composition of Experimental Fish and Diets

At the end of the feeding trial, four fish from each aquarium that had been bled for immunological assays were pooled, stored in plastic bags, and kept at −20°C for subsequent determination of whole body proximate composition. After descaling, fish from each aquarium were finely ground in a Hobart meat grinder and analyzed in duplicate for proximate composition following the standard methods (Association of Official Analytical Chemists [AOAC] 1990). Moisture content was determined by drying samples in an oven at 105°C until constant weight was reached. Samples used for dry matter were digested with concentrated nitric acid and incinerated in a muffle furnace at 600°C overnight for measurement of ash contents. Protein was determined by combustion method using a FP-2000 Nitrogen Analyzer (Leco Corporation, St. Joseph, MI, USA). Lipid content of samples was determined by petroleum ether extraction using a Soxtec System (2055 Soxtec Avanti; Foss Tecator, Höganäs, Sweden). Proximate composition of experimental diets was determined in triplicate using the same procedures.

Immunological Assays

At the end of the growth trial, four fish per aquaria were randomly chosen and anesthetized with MS-222 as previously described. Blood samples were collected from the caudal vasculature using non-heparinized tuberculin syringes and allowed to clot at 4°C overnight. Serum samples were collected following centrifugation at 1,000 × g for 10 min and stored at −80°C for subsequent assays for serum protein, total immunoglobulin, lysozyme activity, and spontaneous haemolytic complement activity (SH50).

Serum protein concentration was determined using the modified Biuret method. Total protein reagent (Sigma, Chemical Co., St. Louis, MO, USA) was added to each well of the microtiter plate at 250 µL/well. Then, 5 µL of serum was added to each well. After 30 min incubation at room temperature, the absorbance of the samples was read at 570 nm. Serum total protein concentrations were calculated using bovine serum albumin as an external standard.

Serum total immunoglobulin was determined following the method of Siwicki and Anderson (1993). The assay was based on the measurement of total protein content in serum prior and post precipitating the immunoglobulin molecules using 12% solution of polyethylene glycol. The difference in protein content was considered as total immunoglobulin content.

Serum lysozyme activity was determined by the method of Litwack (1955) as modified by Sankaran and Gurnani (1972). The assay is based on lysis of lysozyme sensitive gram-positive bacterium *Micrococcus lysodeikticus* (Sigma Chemical Co., St. Louis, MO, USA) by the lysozyme present in the serum. Freeze-dried *M. lysodeikticus* suspension (0.25 mg/mL) was prepared immediately before use by dissolving in sodium phosphate buffer (0.04 M Na_2HPO_4, pH 6.0). Serum (15 µL/well in duplicate) from each of the four fish per tank was placed in a microtiter plate and 250 µL of bacterial cell suspension was added to each well. Hen egg white lysozyme was used as an external standard. The initial and final (after 30 min incubation at 37°C) absorbance of the samples was measured at 450 nm. The rate of reduction in absorbance of samples was converted to lysozyme concentration (µg/mL) using the standard curve.

Spontaneous haemolytic complement activity (SH50) was determined using the method reported by Sunyer and Tort (1995) and modified for using in microtiter plates as described in Lim et al. (2009). Briefly, sheep red blood cells (SRBC) in Alsever's solution (Remel Inc., Lanexa, KS,USA) were added to tilapia serum that had been serially diluted in cold phosphate buffered saline (PBS) solution (0.85% PBS, 0.1% gelatin, 0.15mM $CaCl_2$, and 0.5mM $MgCl_2$) in a round bottom microtiter plate. The plates were incubated at room temperature for 1 h with occasional shaking. After incubation, plates were centrifuged at 800 × g (2,000 rpm) for 10 min at 4°C, and the supernatant pipetted into a new microplate (flat-bottom microtiter). Haemolysis was evaluated spectrophotometrically at 415 nm and converted to percent haemolysis based on distilled water controls. The 50% lysis point (SH50) was calculated by linear regression of each serum sample and expressed as the log dilution.

Disease Challenge

S. iniae (ARS 98-60), originally isolated from hybrid striped bass (*Morone chrysops* × *Morone saxatilis*) with natural streptococcal disease and reisolated from experimentally infected Nile tilapia, was used to challenge tilapia by intraperitoneal (IP) injection. The isolate was identified as *S. iniae* by the methodology described by Shoemaker and Klesius (1997). Frozen stock-culture of *S. iniae* was grown in tryptic soy broth (TSB; Difco Laboratories, Sparks, MD, USA) for 24 h at 28°C. The concentration of the culture was adjusted to an optical density of 1.0 at 540 nm using a spectrophotometer to give an estimated *S. iniae* concentration of 1×10^9 colony-forming units (CFU)/mL. After sampling for immunological assays and measurement of proximate body composition, the number of tilapia remaining in the original aquaria was adjusted to 20, and all fish in each aquarium were challenged by IP injection with 100 µL of *S. iniae* culture containing 1×10^5 CFU/mL (10^4 CFU/fish). After injection, the fish were returned to their respective aquaria. Each group of fish continued to be fed twice daily with the same experimental diet that was assigned in the growth trial. Fish were monitored for mortality and dead fish removed and recorded twice daily for 20 days following injection.

Agglutination Antibody Titer

At the end of the challenge trial (day 21), blood samples were collected from the caudal vasculature of four surviving fish and sera collected following centrifugation and stored at −80°C. Agglutinating antibody titer against *S. iniae* in pre- and post-challenge sera was determined by modifying the method of Chen and Light (1994) as described in Yildirim-Aksoy et al. (2007). Formalin-killed *S. iniae* cells were adjusted to an optical density of 0.8 at 540 nm and added to plasma serially diluted in PBS in a 96-round-bottomed microtiter plate and mixed. Positive plasma from a *S. iniae* infected fish and negative (PBS) were used as assay controls. The plates were covered with plastic film and incubated at room temperature for 16 h. The agglutination end point was established as the last serum dilution where cell agglutination was visible after incubation as compared to the positive control. Agglutination titers were reported as $\log_{10}$ of the reciprocal of this serum dilution. Baseline fish, sampled prior to disease challenge, were negative for *S. iniae*.

Statistical Analysis

Data were analyzed by one-way analysis of variance (ANOVA). Duncan's multiple range tests were used to determine differences between treatment means. Differences were considered significant at the $P < 0.05$. All analysis was performed using the SAS program version 9 (SAS Institute Inc., Cary, NY, USA 2001).

RESULTS

Growth Performance

Mean final weight gain, feed intake, feed efficiency ratio (FER), and survival after 12 weeks of feeding with diets containing various levels of BY and prebiotic GB are given in Table 2. Weight gain, feed intake, and survival were not significantly affected by dietary treatment. However, fish fed the diet supplemented with 1% GB had numerically lower weight gain and significantly lower FER than the group fed other diets. There were no significant differences among FER values of fish in other treatments.

Whole Body Proximate Composition

Whole proximate body composition (moisture, protein, lipid, and ash) did not differ among treatments (Table 3). However, fish fed diets supplemented with dried BY or GB tended to accumulate more body lipid than the fish fed the control diet.

Immune Response

Serum protein, total immunoglobulin, and lysozyme activity were unaffected by dietary treatments (Table 4). Serum spontaneous haemolytic complement activity (SH50) of fish fed the 1% BY diet was significantly higher than those of the groups fed the control diet and diets supplemented with 2% BY or

TABLE 2 Mean Final Weight Gain, Dry Matter Feed Intake, Feed Efficiency Ratio (FER), and Survival of Nile Tilapia Fed Diets Containing Various Levels of Dried Brewers Yeast and GroBiotic-A for 12 Weeks[1]

Treatment	Weight gain (g)	Feed intake/fish (DM basis) (g)	FER[2]	Survival (%)
Control	85.28	113.86	0.75^a	95.85
1% Brewers yeast	85.02	112.66	0.76^a	92.50
2% Brewers yeast	84.88	115.41	0.74^a	97.50
1% GroBiotic-A	77.31	119.15	0.66^b	91.68
2% GroBiotic-A	85.39	115.97	0.74^a	91.68
Pooled SEM	2.47	3.60	0.022	4.08

[1]Values are means of four replicates per treatment. Means in the same column with different superscripts are significantly different at $P < 0.05$.
[2]FER = weight gain (g)/dry feed fed (g).

TABLE 3 Whole Body Proximate Composition of Nile Tilapia Fed Diets Containing Different Levels of Brewers Yeast and GroBiotic-A for 12 Weeks[1]

Treatment	Percent (%) wet weight basis			
	Moisture	Protein	Lipid	Ash
Control	73.40	15.02	6.77	3.53
1% Brewers yeast	73.57	15.21	7.17	3.30
2% Brewers yeast	72.68	15.06	7.06	3.68
1% GroBiotic-A	73.02	14.93	7.36	3.51
2% GroBiotic-A	72.96	14.91	7.26	3.55
Pooled SEM	0.46	0.15	0.25	0.15

[1]Values are means of two determinations of pooled samples of four fish per tank and four tanks per treatment. No significant differences were observed among treatment means at $P < 0.05$.

TABLE 4 Mean Serum Protein, Total Immunoglobulin, Lysozyme, Spontaneous Haemolytic Complement (SH50), and Agglutinating Antibody (Ab) Titer to *S. iniae* in Nile Tilapia Fed Diets Containing Different Levels of Brewers Yeast and GroBiotic-A for 12 Weeks[1]

Treatment	Serum protein (mg/mL)	Total immunoglobulin (mg/mL)	Lysozyme (µg/mL)	SH50 (units/mL)	Ab titer[2] ($\log_{10}$)
Control	40.34	2.66	11.69	67.11[b]	1.26
1% Brewers yeast	39.42	2.20	12.70	130.19[a]	1.06
2% Brewers yeast	38.60	2.38	11.45	62.74[b]	1.43
1% GroBiotic-A	39.79	2.01	9.36	89.46[ab]	1.46
2% GroBiotic-A	39.30	2.20	10.51	78.99[b]	0.75
Pooled SEM	0.91	0.56	2.28	12.50	0.43

[1]Values are means of two determinations per fish (except Ab titer), four fish per tank and four tanks per treatment. Means in the same column with different superscripts are significantly different at $P < 0.05$.
[2]Values are means of one determination per fish, four fish per tank and four tanks per treatment, measured at 21 days post-injection challenge.

GB. There were no significant differences among SH50 values of fish fed fish fed the control diet and diets supplemented with 2% BY and 1 or 2% GB. Serum antibody titers against *S. iniae* at 21 days post-challenge were not significantly affected by dietary treatments, although the values were numerically higher in fish fed 2% BY and 1% GB diets.

Bacterial Challenge

The mean number of days at which the first mortality occurred after *S. iniae* challenge and cumulative mortality at day 20 post-challenge with *S. iniae* were not significantly affected by dietary treatments (Table 5; Figure 1). The non-significant differences among these data were due to large variations among replicate values leading to large experimental errors. Numerically, however, fish fed diets supplemented with 1% BY and 2% BY or GB

TABLE 5 Mean Number of Days to First Mortality and Cumulative Mortality of Nile Tilapia at 20 Days Post-Challenge With *Streptococcus iniae*[1]

Treatment	Days to fist mortality	Cumulative mortality (%)
Control	2.25	66.25
1% Brewers yeast	2.25	50.00
2% Brewers yeast	2.00	50.00
1% GroBiotic-A	2.75	58.80
2% GroBiotic-A	2.00	50.00
Pooled SEM	0.27	7.40

[1]Values are means of four replicates per treatment. No differences were observed among treatment means at $P < 0.05$.

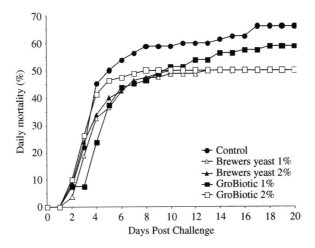

FIGURE 1 Daily cumulative mortality of Nile tilapia after 20 days of challenge with *Streptococcus iniae*.

had substantially lower cumulative mortality (50%) than that of the control treatment (66.25%).

DISCUSSION

There have been numerous studies evaluating the influence of prebiotics such as yeast and its subcomponents (including β-glucans, nucleic acids, and mannan oligosaccharides), as dietary supplements on fish growth performance, immune function, and disease resistance (Duncan & Klesius 1996; Sakai 1999; Gannam & Schrock 2001; Oliva-Teles & Gonçalves 2001; Li & Gatlin 2003, 2004, 2005; Waszkiewicz-Robak & Karwowska 2004; Burr et al. 2005; Whittington et al. 2005; Gatesoupe 2007; Welker et al. 2007; Shelby et al. 2009; Marrifield 2010; Ringø et al. 2010). However, the beneficial effects of dietary prebiotics on growth performance and resistance to infectious diseases are not consistent. Lara-Flores et al. (2003) and Abdel-Tawwab et al. (2008) reported that dietary supplementation of brewers yeast, *S. cerevisiae*, significantly improved growth and feed efficiency of Nile tilapia. In contrast, Li and Gatlin (2003) did not observed significant change in weight gain and feed efficiency of juvenile hybrid striped bass fed diets supplemented with 1%, 2%, or 4% BY. A later seven-week study by the same authors (Li & Gatlin 2004) showed a trend of growth improvement in juvenile fish fed diets supplemented with 1% and 2% BY or GB. However, only fish fed the GB-containing diets had significantly increased feed efficiency. With sub-adults of the same species, Li and Gatlin (2005) obtained generally increased performance of fish fed diets supplemented with 1% and 2% BY and 2% GB at four or six weeks, but significantly enhanced growth and feed efficiency were obtained after 12 and 16 weeks of feeding, respectively. Results of our 12-week feeding study, however, showed that dietary supplementation of BY or GB at 1% or 2% had no effect on weight gain, feed intake, or survival of Nile tilapia, but a significant reduction in feed efficiency was observed in fish fed the 1% GB diet. The decrease in feed efficiency obtained in this treatment cannot be explained but was likely not related to dietary supplementation of GB. Because mixed sex tilapia were used in this study, a higher proportion of females may have been stocked in this treatment. It is a common knowledge that, in tilapia, females grow slower and convert feed less efficiently than males. This is evident as fish in this treatment consumed slightly more feed but gained less weight and accumulated slightly more fat than other groups of fish. Other research showed that the growth and feed efficiency of Nile tilapia (Shelby et al. 2009) and channel catfish (Welker et al. 2007) were unaffected by dietary supplementation of yeast or yeast subcomponents consisted mainly β-glucan or oligosaccharides.

Li and Gatlin (2003) reported no significant changes in whole body composition in juvenile hybrid striped bass fed diets supplemented with 1%, 2%, or 4% BY for eight weeks. In the present study, we also observed no significant differences in whole body composition (moisture, protein, lipid, and ash) of tilapia after 12 weeks of feeding diets supplemented with 1% and 2% BY or GB. However, the group fed the 1% GB diet that consumed more feed, gained less, and converted feed less efficiently tended to accumulate more body lipid. In contrast, Abdel-Tawwab et al. (2008) obtained significant increased in body protein and ash in Nile tilapia fed diets supplemented with 0.1% to 0.5% yeast, respectively. Body total lipid content, however, significantly decreased in fish fed 0.2% and 0.5% yeast diets. They indicated that this alteration in body composition was attributed to increased feed intake, better nutrient utilization, high nutrient digestibility, and increased nutrient deposit.

Even though cumulative mortality of tilapia 20 days following challenge with *S. iniae* was not significantly influenced by dietary treatments, fish fed diets supplemented with 1% BY and 2% BY or GB had substantially lower mortality (50.0%) than those of the groups fed the 1% GB (58.8%) and the control diet (62.3%). Mortality of fish fed diets containing 2% GB or BY and 1% BY ceased 8, 10, and 13 days post-challenge, respectively (Figure 1). For the groups fed the control diet and the diet with 1% GB, mortality continued until day 17 and 18, respectively. Although bacterial count was not performed in this study, the earlier cessation of mortality and lower total mortality in fish fed 1% BY and 2% BY or GB were likely due to lighter infection rate. Li and Gatlin (2003) reported that, after nine weeks of feeding, exposure of juvenile hybrid striped bass to *S. iniae* resulted in reduced signs of disease and no mortality in fish fed 2% and 4% BY diets, while 20% and 10% mortality were obtained in fish fed the control and 1% BY, respectively. In a later study, they obtained significantly enhanced survival after bath exposure with *S. iniae* in hybrid striped bass fed diets with 1% and 2% BY or GB (Li & Gatlin 2004). With sub-adults of the same species, Li and Gatlin (2005) obtained a significant reduction of mortality following *in situ* mycobacterial challenge in fish fed the 2% GB diet relative to the groups fed the control diet and 1% and 2% BY diets at the end of 21 weeks. Results of some earlier studies with prebiotics, particularly yeast and yeast by-products, in fish suggest that their inclusion in diets can enhance the resistance of several fish species against bacterial infections (Raa et al. 1990; Siwicki et al. 1994; Yoshida et al. 1995).

Except for SH50 activity, serum protein, total immunoglobulin, lysozyme activity, and agglutinating antibody titers against *S. iniae* at 21 days post-challenge were unaffected by dietary treatments. The SH50 of fish fed the 1% BY diet was significantly higher than those fed the control and 2% BY or GB diets. These SH50 values did not appear to follow the trend observed for cumulative mortality following *S. iniae* challenge as fish fed diets with 1% BY and 2% BY or GB had the same mortality (50%). Welker et al. (2007) obtained no significant differences in immune function (SH50, lysozyme activity, superoxide anion production, and macrophage bactericidal activity) and resistance to *E. ictaluri* challenge in catfish fed diets supplemented with *S. cerevisiae* or yeast sub-components (glucan and mannan). Duncan and Klesius (1996) reported that channel catfish fed the β-glucan-containing (0.2%) diet had enhanced macrophage and neutrophil migration and phagocytosis, whereas fish fed the *S. cerevisiae*-diet (2.7%) had enhanced phagocytic activity of peritoneal exudate cells. This, however, had no effect on the resistance of fish to *E. ictaluri* infection. Similarly, Ainsworth et al. (1994) reported no improved resistance to *E. ictaluri* challenge in catfish fed 0.1% β-glucan, but obtained increased antibody titers to *E. ictaluri*. In hybrid striped bass, Li and Gatlin (2003) reported improved resistance to *S. iniae* and neutrophil oxidative radical production and extracellular superoxide anion production of head kidney phagocytic cells, but serum lysozyme activity was not affect by dietary inclusion of 1%, 2%, and 4% BY. In later studies, they observed a trend of increasing neutrophil oxidative radical production and intracellular superoxide anion production of head kidney macrophages in fish fed 1% and 2% BY or GB, while extracellular superoxide anion production of head kidney macrophages of fish fed 1% and 2% BY and 1% GB was significantly higher than that of fish fed the control diets. All groups of fish fed BY or GB diets exhibited significantly enhanced survival following immersion challenge with *S. iniae* (Li & Gatlin 2004). In sub-adults of the same species, they obtained significantly improved survival against *in situ* mycobacterial infection in fish fed the 2% GB diet, even though fish in this treatment had significantly lower serum peroxidase and extracellular superoxide anion production of head kidney macrophages obtained (Li & Gatlin 2005). Jorgensen and Robertsen (1995) observed a marked increase in respiratory burst activity of head kidney macrophages of Atlantic salmon four to seven days after treatments with 0.1–1 μg/mL of glucan. Despite the stimulatory effect of glucan on respiratory burst activity, they reported that these macrophages did not show enhanced bactericidal activity against the avirulent or virulent strain of *Aeromonas salmonicida*.

Although our immunological measurements focused on serum components, results of earlier studies have shown that cellular immunity to be more affected by dietary inclusion of prebiotics. However, the effects of dietary immunostimulants on cellular and humoral immune responses as well as the resistance of fish to infectious diseases vary considerably between studies, even for the same immune parameters. Differences in species, fish size, physiological status, quality of diets, source and concentration of prebiotics, feeding duration and levels, challenge method and concentration, and virulence of the pathogens are some of the factors that may have contributed to inconsistencies among various research results.

Data of this study indicated that weight gain, feed intake, survival, and whole body proximate composition of Nile tilapia following 12 weeks of feeding were not significantly affected by dietary supplementation of 1% and 2% BY or GB. The significantly lower feed efficiency observed in fish fed the 1% GB diet was probably not related to dietary treatments. Among the serum immunological components evaluated, only SH50 was significantly affected by dietary treatments. Cumulative mortality 20 days post challenge with *S. iniae* was not significantly affected by dietary treatments. However, fish fed diets supplemented with 1% BY and 2% BY or GB had substantially reduced mortality and mortality ended earlier as compared to those of other treatments.

REFERENCES

Abdel-Tawwab, M., A. M. Abdel-Rahman, and N. E. M. Ismael. 2008. Evaluation of commercial live bakers' yeast, *Saccharomyces cerevisiae* as a growth and immunity promoter for fry Nile tilapia, *Oreochromis niloticus* (L.) challenged *in situ* with *Aeromonas hydrophila*. *Aquaculture* 280:185–189.

Ainsworth, A. J., C. P. Mao, and C. R. Boyle. 1994. Immune response enhancement in channel catfish, *Ictalurus punctatus*, using b-glucan from Schizophyllum commune. In *Modulators of fish immune responses*, vol. 1, edited by J. S. Stolen and T. C. Fletcher, 67–82. Fair Haven, NJ: SOS Publications.

Association of Official Analytical Chemists (AOAC). 1990. *Official methods of analysis*, 15th ed. Arlington, VA: AOAC.

Burr, G., D Gatlin, and S. Ricke. 2005. Microbial ecology of the gastrointtinal tract of fish and the potential application of prebiotics and probiotic in finfish aquaculture. *Journal of the World Aquaculture Society* 36:425–436.

Chen, M. F., and T. S. Light. 1994. Specificity of the channel catfish antibody to *Edwardsiella ictaluri*. *Journal of Aquatic Animal Health* 6:226–270.

Duncan, P. L., and P. H. Klesius. 1996. Dietary immunostimulants enhance nonspecific immune responses in channel catfish but not resistance to *Edwardsiella ictaluri*. *Journal of Aquatic Animal Health* 8:241–248.

Food and Agriculture Organization of the United Nations (FAO). 2002. Antibiotics residue in aquaculture products. In *The State of World Fisheries and Aquaculture 2002*, 74–82 Rome, Italy: FAO.

Fitzsimmons, K. 2006. Prospect and potential for global production. In *Tilapia: Biology, culture and nutrition*, edited by C. Lim and C. D. Webster, 51–72. New York: The Haworth Press.

Gannam, A. L., and R. M. Schrock. 2001. Immunostimulants in fish diets. In *Nutrition and fish health*, edited by C. Lim and C. D. Webster, 235–266. New York: Food Product Press.

Gatesoupe, F.-J. 2007. Live yeasts in the gut: Natural occurrence, dietary introduction, and their effect on fish health and development. *Aquaculture* 267:20–30.

Jorgensen, J. B., and B. Robertsen. 1995. Yeast β-glucan stimulates respiratory bust activity of Atlantic salmon (*Salmo salar* L.) macrophages. *Developmental and Comparative Immunology* 19:43–57.

Lara-Flores, M., M. A. Olvera-Novoa, B. E. Guzmán-Méndez, and W. López-Madrid. 2003. Use of *Streptococcus faecium* and *Lactobacillus acidophilus*, and the yeast *Saccharomyces cerevisiae* as growth promoters in Nile tilapia (*Oreochromis niloticus*). *Aquaculture* 216:193–201.

Li, P., and D. M. Gatlin. 2003. Evaluation of brewers yeast (*Saccharomyces cerevisiae*) as a feed supplement for hybrid striped bass (*Morone chrysops x M. saxatilis*). *Aquaculture* 219:681–692.

Li, P., and D. M. Gatlin. 2004. Dietary brewers yeast and the prebiotic GroBiotic AE influence growth performance, immune responses and resistance of hybrid striped bass (*Morone chrysops x M. saxatilis*) to *Streptococcus iniae* infection. *Aquaculture* 231:445–456.

Li, P., and D. M. Gatlin. 2005. Evaluation of the prebiotic GroBiotic-A and brewers yeast as dietary supplements for sub-adult hybrid striped bass (*Morone chrysops x M. saxatilis*) challenged in situ with *Mycobacterium marinum*. *Aquaculture* 248:197–205.

Lim, C., and C. D. Webster. 2006. *Tilapia biology, culture, and nutrition*. Binghamton, NY: The Haworth Press.

Lim, C., M. Yildirim-Aksoy, and P. H. Klesius. 2009. Growth response and resistance to *Edwardsiella ictaluri* of channel catfish (*Ictalurus punctatus*) fed diets containing distiller's dried grains with soluble. *Journal of the World Aquaculture Society*. 40:182–192.

Litwack, G. 1995. Photometric determination of lysozyme activity. *Proceedings of the Society for Experimental Biology and Medicine* 9:401–403.

Marrifield, D. L., A. Dimitroglou, A. Foey, S. J. Davis, R. T. M. Baker, J. Bøgwald, M. Castex, and E. Ringø. 2010. The current status and future focus of probiotic and prebiotic applications for salmonids. *Aquaculture* 302:1–18.

Muzquiz, J. L, F. M. Royo, C. Ortega, I. De Blias, I. Ruiz, and J. L. Alonso. 1999. Pathogenicity of streptococcosis in rainbow trout (*Oncorhynchus mykiss*): dependence on age of diseased fish. *Bulletin of the European Association of Fish Pathologists* 19:114–119.

National Research Council (NRC). 1993. *Nutrient requirements of fish*. Washington DC: National Academy Press.

Oliva-Teles, A., and P. Gonçalves. 2001. Partial replacement of fishmeal by brewers yeast *Saccaromyces cerevisae* in diets for sea bass *Dicentrarchus labrax* juveniles. *Aquaculture* 202:269–278.

Peres, H., C. Lim, and P. H. Klesius. 2003. Nutritional value of heat-treated soybean meal for channel catfish (*Ictalurus punctatus*). *Aquaculture* 225:67–82.

Raa, J., G. Roerstad, R. Engstad, and B. Robertsen. 1990. The use of immunostimulants to increase resistance of aquatic organisms to microbial infections. In *Diseases in Asian Aquaculture*, edited by I. M. Shariff, R. P. Subasinghe and J. R. Arthur, 39–50. Manila, Philippines: Fish Health Section, Asian Fisheries Society.

Ringø, E, R. E. Olsen, T. Ø. Gifstad, R. A. Dalmo, H. Amlund, G.-I. Hemre, and A. M. Bakke. 2010. Prebiotics in aquaculture: A review. *Aquaculture Nutrition* 16:117–136

Sakai, M. 1999. Current research status of fish immunostimulants. *Aquaculture* 172:63–92.

Sankaran, K., and S. Gurnani. 1972. On the variation in catalytic activity of lysozyme in fishes. *Indian Journal of the Experimental Biochemistry and Physiology* 9:162–165.

Shelby, R. A., C. Lim, M. Yildirim-Aksoy, T. L. Welker, and P. H. Klesius. 2009. Effects of yeast oligosaccharide diet supplements on growth and disease resistance in juvenile Nile tilapia, *Oreochromis niloticus. Journal of Applied Aquaculture* 21:61–71.

Shoemaker, C. A, and P. H. Klesius. 1997. Streptococcal disease problems and control: A review. In *Tilapia aquaculture, vol. 2, Northeast Regional Aquaculture Engineering Service (NRAES) 106*, edited by K. Fitzsimmons, 671–680. Ithaca, NY: NRAES.

Shoemaker, C. A., P. H. Klesius, and J. J. Evans. 2001. Prevalence of *Streptococcus iniae* in tilapia, hybrid striped bass, and channel catfish on commercial fish farms in the United States. *American Journal of Veterinary Research* 62:174–177.

Siwicki, A. K., and D. P. Anderson. 1993. Nonspecific defense mechanisms assay in fish: II. Potential killing activity of neutrophils and macrophages, lysozyme activity in serum and organs and total immunoglobulin level in serum. In

Disease diagnosis and prevention methods, edited by A. K. Siwicki, D. P. Anderson, and J. Waluga, 105–112. Olsztyn, Poland: Stanisław Sakowicz Inland Fisheries Institute (IFI).

Siwicki, A. K., D. P. Anderson, and G. L. Rumsey. 1994. Dietary intake of immunostimulants by rainbow trout affects non-specific immunity and protection against furunculosis. *Veterinary Immunology and Immunopathology* 41:125–139.

Sunyer, J. O., and L. Tort. 1995. Natural hemolytic and bactericidal activities of sea bream *Sparus aurata* serum are affected by the alternative complement pathway. *Veterinary Immunology and Immunopathology* 45:333–345.

Waszkiewicz-Robak, B., and W. Karwowska. 2004. Brewer's yeast as an ingredient enhancing immunity. *Polish Journal of Food and Nutrition Sciences* 13:85–87.

Yildirim-Aksoy M., R. Shelby, C. Lim, and P. H. Klesius. 2007. Growth performance and proximate fatty acid composition of channel catfish, *Ictalurus punctatus*, fed for different duration with a commercial diet supplemented with various levels of menhaden fish oil. *Journal of the World Aquaculture Society* 38:461–474.

Yoshida, T., R. Kruger, and V. Inglis. 1995. Augmentation of nonspecific protection in African catfish, *Clarias gariepinus* (Burchell), by the long-term oral administration of immunostimulants. *Journal of Fish Diseases* 18:195–198.

Welker, T. L., C. Lim, M. Yildirim-Aksoy, R. Shelby, and P. H. Klesius. 2007. Immune response and resistance to stress and *Edwardsiella ictaluri* challenge in channel catfish, *Ictalurus punctatus*, fed diets containing commercial whole cell yeast or yeast subcomponents. *Journal of the World Aquaculture Society* 38:24–35.

Whittington, R., C. Lim, and P. H. Klesius. 2005. Effect of dietary β-glucan levels on the growth response and efficacy of *Streptococcus iniae* vaccine in Nile tilapia, *Oreochromis niloticus*. *Aquaculture* 248:217–225.

Accumulation and Clearance of Orally Administered Erythromycin in Adult Nile Tilapia (*Oreochromis niloticus*) and Giant Freshwater Prawn (*Macrobrachium rosenbergii*)

N. P. MINH[1], T. B. LAM[1], N. T. GIAO[2], and N. C. QUAN[2]

[1]*Department of Food Technology, Ho Chi Minh City University of Technology, Ho Chi Minh City, Vietnam*
[2]*Vietnam-Russia Tropical Center, South Branch, Ho Chi Minh City, Vietnam*

Nile tilapia (Oreochromis niloticus) *and giant freshwater prawns* (Macrobrachium rosenbergii) *were medicated with erythromycin base via a medicated ration at 50 mg and 100 mg erythromycin·kg^{-1} fish body $weight^{-1}·d^{-1}$ for seven days. Erythromycin residues in muscle were determined by LC-MS/MS. After 23 days, erythromycin A residues in the 50 mg treatment were 34.7 ± 9.6 µg/kg in tilapia and 2.8 ± 0.8 µg/kg, and in the 100 mg treatment were 42.9 ± 17.4 µg/kg in tilapia and 31.4 ± 7.5 µg/kg in prawn. Interpolation following European Agency for the Evaluation of Medicinal Products guidelines predicted withdrawal time in tilapia was 33 days (908°C/day) at 50 mg or 42 days (1,150°C/day) at 100 mg under our experimental conditions. After 23 days of withdrawal, erythromycin A residues in prawn tissue were within acceptable limits at a dose of 50 mg, while treatment with 100 mg of erythromycin required a withdrawal time of 35 days (976°C-days), estimated through interpolation. Erythromycin derivatives appeared in the post-dosing stage, but were all depleted to a safe level within the normal withdrawal period.*

The technical assistance of Dr. Nguyen Ba Hoai Anh was highly appreciated.

INTRODUCTION

Streptococcosis, gram positive bacterium, has become a major problem for tilapia farmers and there is still no effective commercial vaccine available. *Streptococcosis* can cause mass death in tilapia farms and, unlike many other tilapia diseases, it will affect even large and otherwise healthy fish. The macrolide antibiotic erythromycin has long been the chemotherapeutant of choice to prevent and control *Streptococcus sp*. Freshwater prawns are also subject to a number of maladies for which erythromycin is the antibiotic of choice.

According to the World Health Organization, Food and Agriculture Organization, the European Union, United States, Canada, Australia, among others, regulate erythromycin residues in seafood for human consumption (Table 1). The Vietnamese Ministry of Agriculture and Rural Development recommends a residual limit of 200 ppb. Farmers can use this substance in their farming provided that they obey the withdrawal time. However, there is limited published research elating to accumulation and clearance of erythromycin in tilapia and adult giant freshwater prawn to guide aquaculturists in its safe use.

The aim of the present study was to follow the uptake, depletion time and derivative metabolism of erythromycin and interpolate a tentative withdrawal time from the edible muscle of Nile tilapia (*Oreochromis niloticus*) and giant freshwater prawn (*Macrobrachium rosenbergii*) after oral administration of the drug in medicated feed.

MATERIALS AND METHODS

One hundred and twenty adult Nile tilapias with an average weight 500 ± 5 g and 750 adult prawns with an average weight 40 ± 2 g were reared on

TABLE 1 Maximum residual limit (MRL, µg/kg) of erythromycin in foodstuffs regulated by importing markets.

Sample	MRL (ppb)					
	Codex (2008)	FAO/WHO (2006)	EU (2008)	US (2005)	Canada (2009)	Australia (2009)
Muscle	100	100	200	100	30	300
Liver	100	100	200	100	30	300
Kidney	100	100	200	100	30	300
Fat	100	100	200	100	30	300
Egg	50	50	150	25	30	300

farm (Ward 7, Soc Trang City, Soc Trang Province, Vietnam, for the tilapia and An Phu Hamlet, Ke Sach District, Soc Trang Province, Vietnam, for the prawns) at pH 6.5–8.5 and O_2 concentration of 5–6 mg.liter^{-1}, and an average temperature of 28 ± 0.5°C and no measurable H_2S, NH_3, NO_2 or CH_4.

The experimental animals were divided into groups A (60 tilapias, 375 prawns) and B (60 tilapias, 375 prawns). Erythromycin base in white powder and purity 96.5% was purchased from DHG Pharma (Can Tho, Vietnam), screened to ensure the absence of derivatives and incorporated into commercially available feeds. The A groups were treated with 50 mg.kg^{-1} body weight day^{-1} for seven days through medicated feed (water temperature, 28°C), while the B groups were treated with 100 mg/kg^{-1} body weight day^{-1} for seven days (water temperature, 28°C).

Muscle samples from five animals of each species at 1, 3, 6, 9, and 23 days post-dosing were collected on farm, placed into polyethylene bags, coded, transferred to the laboratory on dry ice and stored at −40°C prior to analysis for erythromycin A residues based on LC-MS/MS method via TUV Rheinland Aimex Vietnam Co. Ltd., certified by DIN EN ISO/IEC 17025:2005 from DGA (German accreditation). Similarly, biotransformation of erythromycin was monitored by screening and confirming derivative forms of erythromycin at the beginning and end of the sampling period.

RESULTS AND DISCUSSION

Results of erythromycin A depletion at different times in tilapia and prawn samples treated with 50 and 100 mg/kg^{-1} fish body weight/day^{-1} for seven days were shown in Tables 2 and 3, respectively, and depicted graphically by species in Figures 1 and 2. Degree-days are calculated by multiplying the mean daily water temperatures by the total number of days measured.

TABLE 2 Erythromycin depletion at different times in Nile tilapia and giant freshwater prawn muscles treated with 50 mg/kg^{-1} prawn body weight/day^{-1} for seven days.

| Time | | Erythromycin residue in tilapia (μg/kg)^a | Erythromycin residue in prawn (μg/kg)^a |
Day	°C/day		
1	28	22,216.0 ± 22,023.0	15.4 ± 3.3
3	84	13,590.0 ± 14,415.9	10.6 ± 2.1
6	168	940.8 ± 460.3	5.9 ± 3.1
9	252	131.4 ± 31.9	5.5 ± 4.1
23	644	34.7 ± 9.6	2.8 ± 0.8

^{a}Values shown are concentration means ± standard deviations from five tissue samples.

TABLE 3 Erythromycin A clearance at different times in Nile tilapia and giant freshwater prawn fillet samples treated with 100 mg/kg^{-1} prawn body weight/day^{-1} for seven days.

Time		Erythromycin residue in tilapia (μg/kg)^a	Erythromycin residue in prawn (μg/kg)^a
Day	°C/day		
1	28	46,960.0 ± 9,054.7	632.4 ± 74.1
3	84	14,328.0 ± 18,336.1	199.0 ± 31.2
6	168	6,382.0 ± 5,582.5	141.8 ± 3.1
9	252	379.7 ± 99.3	54.2 ± 9.0
23	644	42.9 ± 17.4	31.4 ± 7.5

^{a}Values shown are concentration means ± standard deviations from five tilapia fillet samples.

TABLE 4 Biotranformative forms of erythromycin at 23 days post administration in tilapia and giant prawn muscle samples treated with 50 mg/kg^{-1} body weight/day^{-1} for seven days.

Name of sample	Identification	Test parameter	MDL (μg/kg)	Result (μg/kg)
Erythromycin base	EBS/0901-RC-002	Erythromycin B	1.0	N.D
		Erythromycin C	1.0	N.D
		Erythromycin D	1.0	N.D
		Erythromycin E	1.0	N.D
		Erythromycin F	1.0	5.00
Tilapia muscle	TL-S1	Erythromycin B	10.0	N.D
		Erythromycin C	10.0	N.D
		Erythromycin D	10.0	N.D
		Erythromycin E	10.0	N.D
		Erythromycin F	10.0	N.D
	TL-S5	Erythromycin B	10.0	N.D
		Erythromycin C	10.0	N.D
		Erythromycin D	10.0	N.D
		Erythromycin E	10.0	0.30
		Erythromycin F	10.0	1.37
Giant prawn muscle	GP-S1	Erythromycin B	10.0	N.D
		Erythromycin C	10.0	N.D
		Erythromycin D	10.0	N.D
		Erythromycin E	10.0	N.D
		Erythromycin F	10.0	N.D
Giant prawn muscle	GP-S5	Erythromycin B	10.0	N.D
		Erythromycin C	10.0	N.D
		Erythromycin D	10.0	N.D
		Erythromycin E	10.0	N.D
		Erythromycin F	10.0	N.D

Residual erythromycin A in prawns fed at a rate of 50 mg/kg were within safe limits by day 23. To estimate withdrawal time beyond the end of the 23 days experimental period, data were displayed on a semilogarithmic graph with erythromycin concentration (μg/kg) on the y-axis and post treatment (degree-days) on the x-axis (Figures 3–5). This graphical procedure was obtained through the statistical program recommended by the

European Agency for the Evaluation of Medicinal Products (EMEA, 2009) and downloadable from the same EMEA website. The maximum residue limit (MRL) value for erythromycin was set at 30 $\mu g/kg^{-1}$, as reported by the Canadian Food Inspection Agency (CFIA) on July 11, 2009. Using this statistical method, a withdrawal time of 908°C-days at 50 mg and 1,150°C-days at 100 mg/kg was estimated. Similarly, a withdrawal time of 976°C-days was interpolated for giant freshwater prawn treated for seven days with 100 mg/kg^{-1} prawn body weight/day^{-1} erythromycin.

Of commonly biotransformed erythromycin derivatives, only erythromycin F (5 μg/kg) was present in the erythromycin base. When tilapias were medicated with erythromycin base at the lower dose (group A), none of the derivatives of erythromycin was detected in muscle at day 1 of post-treatment. At day 23 post-treatment (Tables 4 and 5), erythromycin E (0.30 μg/kg) and erythromycin F (1.37 μg/kg) were not at levels of concern. For tilapias fed the higher dose (group B), two derivatives, erythromycin C (131.5 μg/kg) and erythromycin E (258.3 μg/kg) appeared right after ceasing drug treatment. During medication at 100 mg/kg^{-1} erythromycin A slightly

TABLE 5 Biotranformed forms of erythromycin at 23 days post administration in tilapia and giant prawn muscle samples treated with 100 mg/kg^{-1} prawn body weight/day^{-1} for seven days.

Name of sample	Identification	Test parameter	MDL (μg/kg)	Result (μg/kg)
Erythromycin base	EBS/0901-RC-002	Erythromycin B	1.0	N.D
		Erythromycin C	1.0	N.D
		Erythromycin D	1.0	N.D
		Erythromycin E	1.0	N.D
		Erythromycin F	1.0	5.00
Tilapia muscle	TL-SC1	Erythromycin B	10.0	N.D
		Erythromycin C	10.0	131.49
		Erythromycin D	10.0	N.D
		Erythromycin E	10.0	258.28
		Erythromycin F	10.0	N.D
Tilapia muscle	TL-SC5	Erythromycin B	10.0	N.D
		Erythromycin C	10.0	N.D
		Erythromycin D	10.0	N.D
		Erythromycin E	10.0	6.94
		Erythromycin F	10.0	5.90
Giant prawn muscle	GP-SC1	Erythromycin B	10.0	N.D
		Erythromycin C	10.0	N.D
		Erythromycin D	10.0	N.D
		Erythromycin E	10.0	2.09
		Erythromycin F	10.0	N.D
Giant prawn muscle	GP-SC5	Erythromycin B	10.0	N.D
		Erythromycin C	10.0	N.D
		Erythromycin D	10.0	N.D
		Erythromycin E	10.0	5.81
		Erythromycin F	10.0	3.52

N.D = not detected.

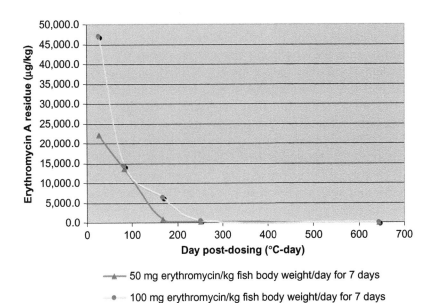

FIGURE 1 Erythromycin A depletion over time in tilapia muscle samples treated with 50 and 100 mg.kg^{-1} fish body weight.day^{-1} for seven days. (Color figure available online.)

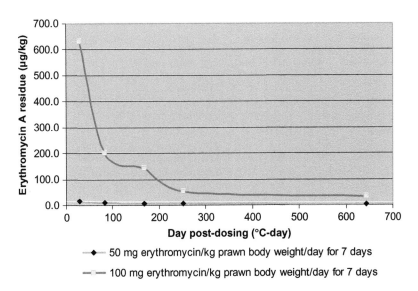

FIGURE 2 Erythromycin A depletion over time in giant prawn muscles treated with 50 and 100 mg/kg^{-1} prawn body weight/day^{-1} for seven days. (Color figure available online.)

changed to erythromycin E (2.09 µg/kg) after one day post-treatment. At day 23 of post-treatment, erythromycin E (0.30 µg/kg) and erythromycin F (1.37 µg/kg) were detected and fortunately were not at levels of concern.

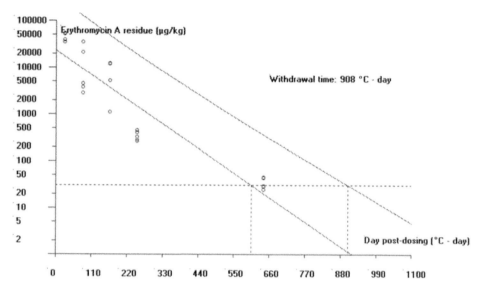

FIGURE 3 Linear regression and upper one-sided tolerance limit (95%) of erythromycin concentrations in tilapia treated with erythromycin for seven days (50 mg/kg^{-1} fish body weight/day^{-1}) versus time. (Color figure available online.)

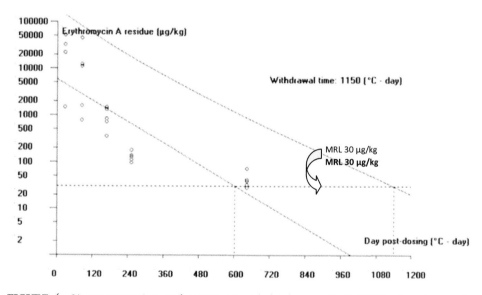

FIGURE 4 Linear regression and upper one-sided tolerance limit (95%) of erythromycin concentrations in tilapia treated with erythromycin for seven days (100 mg/kg^{-1} fish body weight/day^{-1}) versus time. (Color figure available online.)

Our research was designed in conditions that were quite close to actual aquaculture. Based on the results of our study, the mean concentration of erythromycin in group A was lower in comparison with that in group B.

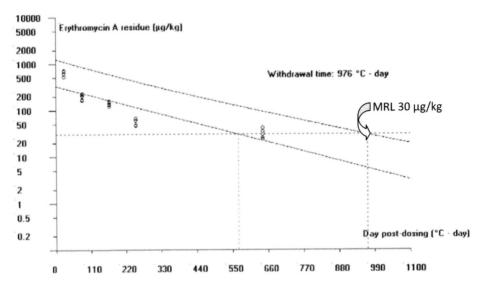

FIGURE 5 Linear regression line and upper one-sided tolerance limit (95%) of erythromycin concentrations in muscle prawn treated with erythromycin for seven days (100 mg/kg^{-1} prawn body weight/day^{-1}) versus time. (Color figure available online.)

However, the eliminating slope of erythromycin residue in group B was faster than in group A.

Drug residue levels dropped quickly during the first three days after treatment termination, then slowly and steadily until a residue level of <100 μg/kg, considered a safe limit by FDA and the European Community requirements, was attained at day 9 of erythromycin withdrawal. However, a longer withdrawal period (35 days of post-treatment) was recommended to ensure complete drug depletion to satisfy the CFIA's concern.

However, it's opportune to underline the fact that, as a general policy, the use of antimicrobials that have an importance in human medicine, like erythromycin, should be limited to the strictly necessary circumstances in veterinary medicine. The potential selection of erythromycin-resistant bacteria in aquaculture settings and the possible dissemination of such resistant clones and/or erythromycin resistance genes to humans might be hazardous for human health.

REFERENCE

EMEA. 2009. *Studies to evaluate the metabolism and residue kinetics of veterinary drugs in food-producing animals: Marker residue depletion studies to establish product withdrawal periods*. European Agency for the Evaluation of Medicinal Products. London, UK. http://www.ema.europa.eu/pdfs/vet/vich/46319909en.pdf

Index

Note: Page numbers in *italic* type refer to tables

Abdel-Tawwab, M. 112–22; *et al* 207
Abdul-Rahman, S.: *et al* 77–95
acclimation 105
Ackman, R.: Heras, H. and Mcleod, C. 191
acocil (*Cambarellus montezumae*) 4, 135–41, 144
adaptation 24–5
Aeromonas hydrophila 112–20
Aeromonas salmonicida 209
Africa 3, 11, 17, 35, 45–6, 50, 60–3
African arowana (*Heterotis niloticus*) 22, 25
African sharptooth catfish (*Clarias gariepinus*) 49, 57, 191
agglutination 208
Agricultural Research and Educational Center, Lebanon (AREC) 79
agriculture 4–13, 17–24, 29–30, 56, 60–1, 92; by-products 49–53, 57, 190; integrated 66, 72, 77–92; production 54–6, 78; recycling 176, 181
agronomy 51
Ahvenharju, T.: and Ruohonen, K. 140
Ainsworth, A.J.: *et al* 208
Akosombo Dam 35
Alaminos, J.: and Domingues, P. 138
Albrektsen, S.: and Torrissen, O. 107
alkalinity 67–8, 74, 105, *106*, 154–5, 174, 185–6
Allan, G.L.: Fielder, D.S. and Bardsley, W.J. 104
Altinok, I.: and Grizzle, J. 109
American ginseng (*Panax quinquefolius*) 112–20
American Public Health Association (APHA) 185
American Tilapia Association 198
American University of Beirut (AUB) 3, 79
ammonia 68, 91, 94–8, 106, 115, 154, 163; -nitrogen 67, 79, 83, 86, 137
ammonium 105; acetate 82; sulfate 74
Anabas testudineus (climbing perch) 183
Anastacido, P.: Oliveira, R. and Marcal, A. 139
Anderson, D.P.: and Siwicki, A.K. 202
anoxia 25
Aquaculture: The Farming and Husbandry of Freshwater and Marine Organisms (Bardach *et al*) 1

Aquaculture (1972) 1
Aquaculture Research and Development Centre, Ghana (ARDEC) 37
aquariums 114–15, 154, 199–203
Arapaima gigas (pirarucu/arapaima) 3, 123–5
Arctic char (*Salvelinus alpinus*) 107
Argentina 153, 161
Arredondo-Figueroa, J. 4; *et al* 135–42; Rodriguez-Serna, M. and Carmona-Osalde, C. 143–51
Artemia (Brine shrimp) 154, 163
Asia 2, 3, 4, 173
Asian stinging catfish (*Heteropneustes fossilis*) 183
Asmah, Hon G. (Minister of Fisheries) 35
Association of Official Analytical Chemists (AOAC, 1990) 114–15, 192
Atlantic salmon (*Salmo salar*) 35, 157–8, 191
Atse, C. 107; Audet, C. and Noue, J. 107
Australasian snapper (*Pagrus auratus*) 104
average daily growth rate (ADGR) 96–100
ayu (*Plecoglossus altivelis*) 158

Balarin, J.D.: and Haller, R.D. 46, 89
Bangladesh 13, 74
Bardsley, W.J.: Allan, G.L. and Fielder, D.S. 104
Barlow, C.G.: and Williams, K.C. 194
barramundi (*Lates calcarifer*) 104
Bautista, M.N. 194
behaviour 165–6
Bekaa plain 77–9, 89
Belal, I.: and Rasheed, N. 3, 96–102
Benfey, T.J. 156–7
Benin 3, 16–31
Bertholletia excelsa (Brazil nut) 131
Bhatt, B.P.: and Bujarbaruah, K.M. 175
Bhujel, R.C.: Pham, T.A. and Little, D.C. 46
Bierbach, D.: Linsenmair, K. and Hauber, M. 16–33
biodiversity 182, 188
biofiltration 163
biomass 25–8, 89–90, 124, 135–6, 140, 161, 165–6
birds 173–4, 184

INDEX

blood meal 3, 123–31
Blue Revolution 16
bluntnose black bream (*Megalobrama amblycephala* Yih) 129
body weight increase (BWI) 105–7
Boyd, C.E. 67; Diana, J.S. and Seim, W.K. 91
Brackishwater Aquaculture Research Centre, Malaysia 192
Braun, N.: *et al* 154–6
Brazil 3, 124, 153, 163–4
Brazil nut (*Bertholletia excelsa*) 131
breeding 19, 53; selective 3, 63
Breine, J.J.: *et al* 66
Brevimyrus niger 25
brewers yeast (BY) 197–209
brine shrimp (*Artemia*) 154, 163
broodfish/broodstock 103–9, 162
brook trout (*Salvelinus fontinalis*) 157–8
Brown, P.B. 144
Brummett, R. 1–4; Etaba Angoni, D. and Pouomogne, V. 62; Gatchouko, M. and Pouomogne, V. 49–64
Bujarbaruah, K.M.: and Bhatt, B.P. 175
Bureau, D.: and El-Haroum, E. 131; Guo, J. and Wang, Y. 131
Bureau of Fisheries and Aquatic Resources - National Freshwater Fisheries Technology Center (BFAR-NFFTC) 97

cage aquaculture 34–47
Caisse Local de Crédit Agricole Mutuel (CLCAM) 29
calcium 109, 174–5, 190
Cambarellus montezumae (acocil) 4, 135–41, 144
Cambaridae (freshwater crayfish) 136, 144
Cameroon 3, 49–63
Canadian Food Inspection Agency (CFIA) 217, 220
canals 66, 183
cannibalism 138–9, 145, 165
Capblancq, J.: and Dauta, A. 67
capital 6, 30, 61
capture fishery 35, 50
Carassius gibelio (gibel carp) 131
Carmona-Osalde, C.: Arredondo-Figueroa, J. and Rodriguez-Serna, M. 143–51; *et al* 139; Olvera-Novoa, M. and Rodriguez-Serna, M. 147
Catacutan, M.R.: and Coloso, R.M. 194
catfish (*Siluriformes*) 50, 152–8, 161–2, 166, 207–8; *Clarias* 22, 25–30, 49, 57, 183, 191
Catla catla (Indian carp) 173–4, 184
cattle 21, 173–6
Cavero, B.A.S.: *et al* 124
Central Laboratory for Aquaculture Research, Egypt 113–15
Central Mexican Plateau 135–6

Centre Communale pour le Promotion Agricole (CeCCPA) 18
channel catfish (*Ictalurus punctatus*) 207
Chen, M.F.: and Light, T.S. 203
Cherax destructor (common yabby) 144
Cherax quadricarinatus (redclaw crayfish) 139
Cherax tenuimanus (marron) 144
chickens 60, 66, 74, 176
Chou, R.: Fuchs, J. and Cuzon, G. 194
Cirrhinus mrigala (Mrigal carp) 171–3, 184
Clargo, M. 12
Clarias 22, 25–30, 49, 57, 183, 191
Clarias batrachus (walking catfish) 183
Clarias gariepinus (African sharptooth catfish) 49, 57, 191
climate 17, 53, 60, 172
climbing perch (*Anabas testudineus*) 183
clove (*Eugenia caryophyllata* thunb) 104–5
Clymo, R.: Ohnstad, M. and Golterman, H. 154
Coche, A.G. 89
Coloso, R.M.: and Catacutan, M.R. 194
Colossoma macropomum (tambaqui) 124–5
common carp (*Cyprinus carpio*) 50, 184
common yabby (*Cherax destructor*) 144
community 60, 182
consumption 17–18, 28–30, 51, 140, 147
consumptive water use (CWU) 80, 90–1
cows *see* cattle
crayfish 137–40, 144–7; freshwater (*Cambaridae*) 136, 144; growth 143–8
crops 3–10, 54, 56, 60, 88–91, 172; production 66, 77–81
crude protein (CP) 143, 199
crustaceans 138–9, 144, 148
Ctenopharyngodon idella (grass carp) 173, 184
cuneate drum (*Nibea miichthioides*) 131
Cuzon, G.: Chou, R. and Fuchs, J. 194
Cyprinus carpio (common carp) 50, 184

daily growth index (DGI) 127
daily weight gain (DWG) 67
dams 35, 50
Dauta, A.: and Capblancq, J. 67
Davies, S.J.: Fasakin, E.A. and Serwata, R.D. 131
decapods 138–40
Dennis, R.: and Thomas, S. 12
Descy, J.P. 67
Deutsche Gesellschaft für Technische Zusammenarbeit (GTZ) 50
development 12, 16–17, 46, 78, 107, 166; projects 49–63; rural 50, 61–2; sustainable 180–8
Dey, M.M.: *et al* 90–1
Diana, C.: and Yi, Y. 89
Diana, J.S.: and Lin, C.K. 89; Seim, W.K. and Boyd, C.E. 91

INDEX

diet 112–20, 140, 143–8; replacement/alternative 123–31, 190–5; supplementary 197–209
diploidy 152–8, 161–6
disease 78, 198–9, 206–9, 214
dissolved oxygen (DO) 67–8, 83–6, 98, 105–6, 137, 152–8, 181
Domingues, P.: and Alaminos, J. 138
Dossou, S. 19
ducks 66, 74, 172–6, 184
Duncan, P.L.: and Klesius, P.H. 208
Dupriez, H.: and De Leener, P. 60
Dytham, C. 116
Dzemeni 34–6, 43–6

ecdysis 138–9
Echinochloa colona 20
ecology 51, 162, 182–4
economics 49–52, 60–1, 78–80, 90; small-scale aquaculture 34–47
economy 17, 23, 35–6, 53, 90–2, 136
ecosystem 46, 180
education 7, 10–11, 49, 54, 55, 184
Edwardsiella ictaluri 208
Edwardsiella tarda 120
effluent 74, 77, 80–91, 181–3, 187
Eichhornia crassipes (common water hyacinth) 20
Eknath, A.E.: *et al* 62
electrical conductivity (EC) 67–8, 105
Eleutherococcus senticosus (Siberian ginseng) 120
environment 1, 124, 136, 153, 181
Epinephelus coioides (orange-spotted grouper) 124
Epinephelus malabaricus (Malabar grouper) 131
Erythromycin 213–20
Espe, M.: Haaland, H. and Njaa, L. 191
Etaba Angoni, D.: Pouomgne, V. and Brummett, R. 62
Eugenia caryophyllata thunb (clove) 104–5
European Agency for Evaluation of Medicinal Products (EMEA) 217
European Aquaculture Society (EAS) 1
European Medicines Agency (EMA) 213
European Patent Office (EPO) 2
European Union (EU) 198, 214
eutrophication 73
Experimental Plant of Aquatic Production (EPAP) 140

Fagbenro, O.: Jauncey, K. and Haylor, G. 191
Fanimo, A.O.: *et al* 193
farming *see* agriculture
Fasakin, E.A.: Serwata, R.D. and Davies, S.J. 131
feed 6, 35, 46, 115–16, 135–41; commercial 143–8; conversion ratio (FCR) 39, 88–9, 96–101, 105–9, 123–7, 137–9, 146–7; efficiency ratio (FER) 126, 201, 204, 207; regime 21–2, 28; replacement/alternative 123–31, 190–5; supplementary 21, 197–209; trial 190, 201

fermented prawn waste liquor (FPWL) 190–5
fertigation 91–2
fertilization 59, 66, 82, 86–91, 105, 154, 163
fertilizer 6, 65–8, 73–4, 77–80, 85, 172–5; inorganic 77, 80, 84, 90; organic 72
Fielder, D.S.: Bardsley, W.J. and Allan, G.L. 104
fingerlings 22, 29–30, 42–3, 61–7, 97–100, 113, 199
fisheries 16–17, 22, 37, 50, 61, 180, 186–90
fishmeal 4, 125, 190–5, 193
flooding 19–27, 30, 46
floodplain 18, 22, 27–30, 124
floriculture 183
Food and Agriculture Organization (FAO), UN 1, 50, 63, 214
Food and Drug Administration (FDA) 220
Franco, L. 74
freshwater crayfish (*Cambaridae*) 136, 144
freshwater fish 103–9, 124, 135, 161–3, 181, 185, 198
Freshwater Fish Biology and Fish Culture Laboratory, Brazil 163
Fuchs, J.: Cuzon, G. and Chou, R. 194
Fukushima, H.: *et al* 161–70

Ganges River 180–2
Gatchouko, M.: Pouomgne, V. and Brummett, R. 49–64
Gatlin, D.: and Li, P. 207–8
Genetically Improved Farmed Tilapia (GIFT) 89, 97, 101
Ghana 3, 27, 34–47
Ghana Agro Food Company Limited (GAFCO) 37
giant freshwater/river prawn (*Macrobrachium rosenbergii*) 213–20
giant tiger prawn (*Penaeus monodon*) 191
gibel carp (*Carassius gibelio*) 131
gilthead sea bream (*Sparus aurata*) 129–31
ginsenosides 118–20
goats 60, 74, 172–6
Goda, A.M.A. 116–18
Golterman, H.: Clymo, R. and Ohnstad, M. 154
Gomes, J.S.: and Olivia-Teles, A. 124
Gooley, G.J.: Ingram, B.A. and McKinnon, L.J. 104
Gorissen, G. 43
grass carp (*Ctenopharyngodon idella*) 173, 184
Green, B.W.: and Teichert-Coddington, D.R. 46
Grizzle, J.: and Altinok, I. 109
GroBiotic-A (GB) 197–209
growth 41, 104, 109, 152, 161–2, 166–8, 171; density, effect 135–41; *Oreochromis niloticus* (Nile tilapia) 65–74, 77, 82–3, 96–101, 197–209; performance 88, 112–20, 129–31, 143–5, 197–209; rate 41–6, 89, 113, 123–4, 140, 144–8, 198; sea bass 190–5
Guerrero, R.D. 89

INDEX

Guo, J.: Wang, Y. and Bureau, D. 131
Gurnani, S.: and Sankaran, K. 202

Haaland, H.: Njaa, L. and Espe, M. 191
habitation 17, 124
haemolytic complement activity (SH50) 202
Haller, R.D.: and Balarin, J.D. 46, 89
El-Haroum, E.: and Bureau, D. 131
harvest 22–8, 42–3, 58, 65–7, 80–2, 86–8, 137
hatcheries 3, 61–3, 100, 105, 125, 163
Hauber, M. 3; Bierbach, D. and Linsenmair, K. 16–33
Haylor, G.: Fagbenro, O. and Jauncey, K. 191
health 9, 54, 220
Hepher, B. 39
Heras, H.: Mcleod, C. and Ackman, R. 191
Heterobranchus longifilis (vundu) 25
Heteropneustes fossilis (Asian stinging catfish) 183
Heterotis niloticus (African arowana) 22, 25
Himalayas 171–8
Hooghly River 183
Hopkins, K.D.: *et al* 90
horticulture 181–3
Huergo, G.: and Zaniboni-Filho, E. 154, 162–3
humidity 79, 171–3
husbandry: animal 10, 145–8, 174–6, 183, 199
Hussain, G.: and Al-Jaloud, A. 91
Hussein, A.H.A. 91–2
hybrid striped bass (*Morone saxatilis* X *Morone chrysops*) 203, 207–8
hydrography 161
hydrology 19
Hypophthalmichthys molitrix (silver carp) 171–6, 184
hypoxia 152, 156–8

Ictalurus punctatus (channel catfish) 207
income 6–10, 18–19, 29, 54, 55, 78
incubation 105
India 4, 74, 171–80
Indian carp (*Catla catla*) 173–4, 184
Indian Council of Agricultural Research (ICAR) 172
individual food consumption (IFC) 135–7, 140–1, 146
Indo-Gangetic Plain 187–8
Indoor Aquaculture Recirculation Facility, UAE 97
industrialization 2, 62, 181
Ingram, B.A.: McKinnon, L.J. and Gooley, G.J. 104
instantaneous rate of increase (IRI) 137
integrated agriculture-aquaculture (IAA) 66, 72, 77–92, 171–8
International Development Research Centre (IDRC) 50

International Fund for Agricultural Development (IFAD) 50
irrigation 10, 19, 77–92
Iwama, G.K.: and Morgan, J.D. 109
Al-Jaloud, A.: and Hussain, G. 91

Jauncey, K.: Haylor, G. and Fagbenro, O. 191
Jones, D.A.: *et al* 144
Jorgensen, J.: and Robertsen, B. 209
Journal of Applied Aquaculture 3–4
jundiá (*Rhamdia quelen*) 3, 152–8, 161–8

Kamaruzzaman, N.: *et al* 46
Kanangire, C.: and Léonard, V. 67
Katerji, N.: Mastrorilii, M. and Rana, G. 91
Ketola, H.G.: Luzier, M.J. and Summerfelt, R.C. 129
Kihongo, V.: and Wetengere, K. 3–15
Klesius, P.H.: and Duncan, P.L. 208; and Shoemaker, C.A. 203
Kolkata 4, 180–8
Kolkata Port Trust (KPT) 182
Kozak, P.: *et al* 139
Kpong Dam 35
Kudi, T.M.: *et al* 6, 10
Kuria labeo (*Labeo gonius*) 173

Labeo bata 184
Labeo gonius (Kuria labeo) 173
Labeo rohita (rohu) 173–4, 184
labour 6, 23, 39, 61
Lactobacillus 191
lakes 3, 34–47, 124, 136
land 19, 29, 49, 52, 55
Landergren, P. 109; and Vallin, L. 107
Lara-Flores, M.: *et al* 207
larviculture 103–9, 154, 161–8
Lates calcarifer (barramundi) 104
Lebanon 3, 77–9, 91
Lebas, F.: *et al* 66
Leener, P. De: and Dupriez, H. 60
Léonard, V.: and Kanangire, C. 67
lethal concentration 50 (LC50) 152–6
Lévêque, C.: *et al* 18
Li, P.: and Gatlin, D. 207–8
Light, T.S.: and Chen, M.F. 203
Lim, C.: *et al* 202
Lin, C.: and Suresh, A. 89
Lin, C.K.: and Diana, J.S. 89
Linsenmair, K.: Hauber, M. and Bierbach, D. 16–33
lipids 145, 197, 200–1, 204, 207
Litopenaeus vannamei (whiteleg shrimp) 120
Little, D.C.: Bhujel, R.C. and Pham, T.A. 46
Litwack, G. 202
Liu, X.L.: *et al* 120
livestock 66, 172–6, 191

INDEX

Lowe-McConnell, R.H. 27–8
Ludwigia 20
Luzier, M.J.: Summerfelt, R.C. and Ketola, H.G. 129
Lymbery, A.J.: and Partridge, G.J. 104
lysozyme 198, 202–4, 208

McClain, W.R. 140
McCrosckey, R. 66
Maccullochella peelii (Murray cod) 129
Macintosh, D.J.: *et al* 46
McKinnon, L.J.: Gooley, G.J. and Ingram, B.A. 104
Mcleod, C.: Ackman, R. and Heras, H. 191
MacLeod, M.G. 109
Macrobrachium rosenbergii (giant freshwater/river prawn) 213–20
Mair, G.C.: *et al* 46
maize (*Zea mays*) 54, 60–1, 77, 81–2, 86–8, 91–2
Maja brachydactyla (spider crab) 138
Malabar grouper (*Epinephelus malabaricus*) 131
Malanville 16–19, 27–30
Malaysia 4, 190–2
Mamun, S.M.: *et al* 101
mango tilapia (*Sarotherodon galilaeus*) 22, 25
Manikhal canal 183
Marcal, A.: Anastacido, P. and Oliveira, R. 139
marron (*Cherax tenuimanus*) 144
Martínez-Llorens, S.: *et al* 129
Masawe, J.L. 6, 10–12
Maselle, A.E.: Misanyiwa, Z.S. and Namwata, B.M. 8
Mastrorilii, M.: Rana, G. and Katerji, N. 91
maximum residue limit (MRL) 217
medicine 113, 220
Megalobrama amblycephala (bluntnose black bream) 129
meiosis 162
Mexico 4, 135–6, 143–4, 153, 161
Micrococcus lysodeikticus 202
Mikalitsa, S.M. 6, 12
Minh, N.P.: *et al* 213–20
Ministry of Agriculture and Rural Development, Vietnam (MARD) 214
Misanyiwa, Z.S.: Namwata, B.M. and Maselle, A.E. 8
Mohammadi, M.: *et al* 3, 103–11
Mohammed-Suhaimee, A.: Zakaria, Z. and Ng, W. 194
monocropping 78
Morgan, J.D.: and Iwama, G.K. 109
Moritz, T.: *et al* 18
Morogoro 5–13
Morone chrysops x Morone saxatilis (hybrid striped bass) 203, 207–8
mortality 34–7, 43–6, 115–16, 138–9, 153–8, 161–4, 203–9

Mozambique tilapia (*Oreochromis mossambicus*) 184
Mrigal carp (*Cirrhinus mrigala*) 171–3, 184
Mudialy Fishermen Cooperative Society (MFCS) 180–8
Mugivane, F.I.: Otieno, P.S. and Nyikal, R.A. 12
Murray cod (*Maccullochella peelii*) 129
Mycobacterium marinum 199

Nafisi, B.M. 104
Namwata, B.M.: Maselle, A.E. and Misanyiwa, Z.S. 8
Nana, J-P.: Pouomegne, J. and Pouomogne, V. 58
National Fish Culture Service, Rwanda (SPN) 66
National Institute of Amazonian Research (INPA) 125
National Research Council (NRC) 200
Nauplii (crustacean larvae) 154, 163
Neptunia oleracea (water mimosa) 20
Ng, W.: Mohammed-Suhaimee, A. and Zakaria, Z. 194
Nibea miichthioides (cuneate drum) 131
Niger River 16–19, 29
Nigeria 10, 28
Nile tilapia (*Oreochromis niloticus*) 49, 57, 62, 79, 183–4, 191; diets 112–20; Erythromycin accumulation 213–20; growth performance 65–74, 77, 82–3, 96–101, 197–209; small-scale cage aquaculture 34–47; *Streptococcus iniae* resistance 197–209
nitrate-nitrogen 67, 73
nitrates 70, 98, 185
nitrite-nitrogen 79, 84
nitrites 67–8, 97–8, 105–6, 126, 154, 163
nitro blue tetrazolium (NBT) 115–18
nitrogen 71–4, 77, 81–2, 85, 88–91, 154, 174–5
Njaa, L.: Espe, M. and Haaland, H. 191
non-governmental organizations (NGOs) 34–5, 182, 188
non-improved strain (NS) 98–101
Nor, N.M. 190–6
Noue, J.: Atse, C. and Audet, C. 107
Noun Division, Western Province 49–54, 57, 60–1
Noun River 53
nutrients 89–91, 127, 131, 174, 183, 186–7, 198–9
nutrition 125, 139, 147, 171–8
Nwanna, L.C. 191–3
Nyikal, R.A.: Mugivane, F.I. and Otieno, P.S. 12

Ofori, J.K.: *et al* 3, 34–48
Ohnstad, M.: Golterman, H. and Clymo, R. 154
oligotrophy 35
olive flounder (*Paralichthys olivaceus*) 120
Oliveira, R.: Marcal, A. and Anastacido, P. 139
Olivia-Teles, A.: and Gomes, J.S. 124
Olsen, S.R.: and Watanabe, F.S. 82

INDEX

Olvera, N.M.A.: et al 145–6
Oncorhynchus masou macrostomus (red-spotted masu salmon) 158
Oncorhynchus mykiss (rainbow trout) 103–9, 144–8, 158
Oncorhynchus (Pacific salmon) 107
orange-spotted grouper (*Epinephelus coioides*) 124
Oreochromis mossambicus (Mozambique tilapia) 184
Oreochromis niloticus (Nile tilapia) 49, 57, 62, 79, 183–4, 191; diets 112–20; Erythromycin accumulation 213–20; growth performance 65–74, 77, 82–3, 96–101, 197–209; small-scall cage aquaculture 34–47; *Streptococcus iniae* resistance 197–209
Organization for Economic Cooperation and Development (OECD) 2
Osteoglossidae 25, 124
Otieno, P.S.: Nyikal, R.A. and Mugivane, F.I. 12
Ouémé River 17, 21, 27, 28

Pacifastacus leniusculus (signal crayfish) 136, 139–40
Pacific salmon (*Oncorhynchus*) 107
Pagrus auratus (Australasian snapper) 104
Panax quinquefolius (American ginseng) 112–20
Paralichthys olivaceus (olive flounder) 120
Partridge, G.J.: and Lymbery, A.J. 104
Paspalum scrobiculatum (millet) 20
Paugy, D.: et al 18
Peace Corps 50
Penaeus monodon (giant tiger prawn) 191
Pereira-Filho, M.: et al 124
Pham, T.A.: Little, D.C. and Bhujel, R.C. 46
Phillips, R.B.: et al 154
phosphate 74, 185; buffered saline (PBS) 202–3
phosphorus 67, 71, 74, 81–2, 85, 88–91, 174–5
physiology 155–8, 166
phytoplankton 74, 174
pigs 66, 143–8, 173–6
pirarucu/arapaima (*Arapaima gigas*) 3, 123–5
Pistia stratiotes (water cabbage/lettuce) 20
plankton 174, 182, 186–7
plants 20, 86–8, 182; growth 80–2, 89; non-fertilized 84
Plecoglossus altivelis (ayu) 158
ploidy 3–4, 152–8, 161–8
Poacea (true grasses) 20
politics 9, 49, 61
pollution 4, 24, 30, 136, 180–2
polyculture 49
Polypterus senegalus (Senegal bichir) 25
polysaccharides 198
ponds 22, 49, 51–3, 57–60, 58, 172–4, 184–7; earthen 50, 65–74
potassium 81–2, 85, 88–91, 109, 174–5
poultry 171–5

Pouomgne, V.: Brummett, R. and Etaba Angoni, D. 62; Brummett, R. and Gatchouko, M. 49–64; Nana, J-P. and Pouomgne, J. 58
poverty 3–8
prawns 4, 193, 215–17
predation 9, 30, 58
Procambarus acutus (white river crayfish) 136, 139–40
Procambarus clarkii (red swamp crayfish) 136, 139–40, 144
Procambarus llamasi 139–40, 143–8
production parameters 34–47
profitability 3–5, 10, 13, 42–3, 60, 90–1
protein 123–31, *145*, 193, 197–204, 207; efficiency ratio (PER) 127, 190
Protopterus annectens (West African lungfish) 25
purification 183–5

Quagrainie, K.K.: et al 10

rabbits 65–74, 143–8
Raemaekers, R.H. 54
Railo, E.: Savolainen, R. and Ruohonen, K. 139
rainbow trout (*Oncorhynchus mykiss*) 103–9, 144–8, 158
Rana, G.: Katerji, N. and Mastrorilii, M. 91
rapid appraisal of agricultural knowledge system (RAAKS) 51
Rasheed, N.: and Belal, I. 3, 96–102
Recirculating Aquaculture System (RAS) 96–101, 143, 148, 163
recycling 181–3
red swamp crayfish (*Procambarus clarkii*) 136, 139–40, 144
red-spotted masu salmon (*Oncorhynchus masou macrostomus*) 158
redclaw crayfish (*Cherax quadricarinatus*) 139
reproduction 96, 103–9, 124, 140
Research Tool for Natural Resource Management, Monitoring and Evaluation (RESTORE) 49–51
Rhamdia quelen (jundia) 152–8, 161–8
Ribeiro, R.A.: et al 123–34
Ridha, M.T. 100–1
rivers 17–19, 30, 35, 46, 124, 180–2
Robertsen, B.: and Jorgensen, J. 209
Rodriguez-Serna, M.: Carmona-Osalde, C. and Arredondo-Figueroa, J. 143–51; Carmona-Osalde, C. and Olvera-Novoa, M. 147; et al 139, 144
rohu (*Labeo rohita*) 173–4, 184
Rook, G.A.W.: et al 115
Ruohonen, K.: and Ahvenharju, T. 140; Railo, E. and Savolainen, R. 139
Rwanda 3, 65–74, 66, 72–3

Saccharomyces cerevisiae 199, 208
salinity 103–9, 115, 185

INDEX

Salmo salar (Atlantic salmon) 35, 157–8, 191, 209
Salmo trutta (sea/brown trout) 107–9
Salvelinus alpinus (Arctic char) 107
Salvelinus fontinalis (brook trout) 157–8
Sankaran, K.: and Gurnani, S. 202
Saoud, I. 3
Sarotherodon galilaeus (mango tilapia) 22, 25
Saudi Arabia 91, 97
Savolainen, R.: Ruohonen, K. and Railo, E. 139
El-Sayed, A.F.M. 46, 101
Schaperclaus, W.: *et al* 115
Scombridae 51
sea bass 190–5
sea/brown trout (*Salmo trutta*) 107–9
Seim, W.K.: Boyd, C.E. and Diana, J.S. 91
selected line (SL) strain 100
Senegal bichir (*Polypterus senegalus*) 25
Sengupta, A. 4; *et al* 180–9
serum haemolytic complement activity (SH50) 198, 204–5, 208
Serwata, R.D.: Davies, S.J. and Fasakin, E.A. 131
Shastri, R.K. 6, 11
sheep red blood cells (SRBC) 202
Shoemaker, C.A.: and Klesius, P.H. 203
shrimp 135, 140, 143–8
Siberian ginseng (*Eleutherococcus senticosus*) 120
signal crayfish (*Pacifastacus leniusculus*) 136, 139–40
Siluriformes (catfish) 50, 152–8, 161–2, 166, 207–8; *Clarias* 22, 25–30, 49, 57, 183, 191
Silva, L.V.F. 165
silver carp (*Hypophthalmichthys molitrix*) 171–6, 184
Siwicki, A.K.: and Anderson, D.P. 202
sodium nitrate 74
sodium phosphate 202
soil 53, 77, 172
Sota river 17–19
South America 124, 163
Sparus aurata (gilthead sea bream) 129–31
specific growth rate (sgr) 39, 80–2, 105–7, 115–16, 137–9, 146, 193–4
spider crab (*Maja brachydactyla*) 138
Ssegane, H. 3
stocking 34–5, 42–3, 81, 83, 99, 100, 154; density 89, 101, 124, 136–8, 144, 161–8, 173
streams 181
Streptococcus 214
Streptococcus iniae 197–209
stress 153, 162, 192
sulfates 105, 185
Summerfelt, R.C.: Ketola, H.G. and Luzier, M.J. 129
Sunyer, J.: and Tort, L. 202
Suresh, A.: and Lin, C. 89
survival 96, 103–4, 109, 152, 155, 161–6, 201–4; rate (SR) 67–8, 82, 139, 168

sustainability 35, 60, 180–8
Synodontis 25

Tabaro, S.R.: *et al* 3, 65–76
tambaqui (*Colossoma macropomum*) 124–5
Tanzania 3, 5–13
Teichert-Coddington, D.R. 74; and Green, B.W. 46
temperature 44, 67, 79, *100*, 126, 183–5, 199; water 83, 86, 115, 153, 158, 174–5, 215
theft 9, 29, 46, 58
Thomas, S.: and Dennis, R. 12
Tilapia mossambica see *Oreochromis mossambicus* sp
Tilapia nilotica see *Oreochromis niloticus* sp
titration 106
Toko, I. 29
Torrissen, O.: and Albrektsen, S. 107
total ammonia-nitrogen (TAN) 79, 83, 86, 137
total dissolved solid (TDS) 105–6
Traite de Pisciculture (Huet) 1
Trewavas, E. 18
triploidy 3–4, 152–8, 161–8
turbidity 44, 65, 74
turkeys 143–8
TUV Rheinland Aimex Vietnam Co Ltd 215

Uganda 3, 11, 35, 46, 62
United Arab Emirates (UAE) 3, 96–101
United Nations Development Programme (UNDP) 50; Food and Agriculture Organization (FAO) 1, 50, 63, 214
United Republic of Tanzania (URT) 8
United States Agency for International Development (USAID) 50
United States of America (USA) 2, 155, 214; Environmental Protection Agency (EPA) 155; Food and Drug Administration (FDA) 220; Patents and Trademarks Office 2
urbanization 181
Uruguay 162
Uruguay River 153

Vallin, L.: and Landergren, P. 107
Vechklang, K.: *et al* 197–212
Vibrio anguillarum 120
Vietnam 4, 214
Village Community Banks (VICOBA) 13
Vivekananda Institute of Medical Sciences 4
Vleet, J. Van 66
Volta Lake 3, 34–47
Volta River 35, 46
voluntary feed intake (VFI) 126–7, 131
vundu (*Heterobranchus longifilis*) 25

walking catfish (*Clarias batrachus*) 183
Wang, Y.: Bureau, D. and Guo, J. 131
waste 183, 190–1, 201

INDEX

wastewater aquaculture 180–8, 184
Watanabe, F.S.: and Olsen, S.R. 82
water 9, 17, 41, 68, 77–8; brackish 103–9; fresh/clean 78, 103, 106, 114, 181; productivity 77–80, 85, 88, 91; quality 39–43, 65–74, 78–9, 100, 163–5, 180–7, 193; temperature 83, 86, 115, 153, 158, 174–5, 215; underground 101–9; use efficiency (WUE) 77–92; value index (WVI) 85, 88–91; well 77, 82–8, 91
water mimosa (*Neptunia oleracea*) 20
Water Research Institute, Ghana (WRI) 35–7, 42, 46
weight gain (wg) 105–7, 126, 164, 190, 201, 204, 207
Weiss, L.: and Zaniboni-Filho, E. 3, 152–60
Welcomme, R.L. 30
Welker, T.L.: *et al* 208
West African lungfish (*Protopterus annectens*) 25
West Bengal 180–8
Wetengere, K. 8; and Kihongo, V. 3–15
whedos (*tschifi dais*) 3, 16–31
white crayfish 136, 139–40
white river crayfish (*Procambarus acutus*) 136, 139–40

whiteleg shrimp (*Litopenaeus vannamei*) 120
Williams, K.C.: and Barlow, C.G. 194
Won, K.M.: *et al* 120
Word Health Organization (WHO) 214
World Aquaculture Society 1
World Bank (WB) 9, 50
WorldFish Center 36, 51

Yi, Y.: and Diana, C. 89
Yildirim-Aksoy, M.: *et al* 203
Yucatan Peninsula 143–4

Zakaria, Z. 4; Ng, W. and Mohammed-Suhaimee, A. 194
Zaniboni-Filho, E.: and Huergo, G. 154, 162–3; and Weiss, L. 3, 152–60
Zea mays (maize) 54, 60–1, 77, 81–2, 86–8, 91–2
Zhou, Z.: *et al* 129
Zimbabwe 35, 62
Zoological Survey, India (1991) 184
zooplankton 174, 180, 185–6